Kenya's Changing Landscape

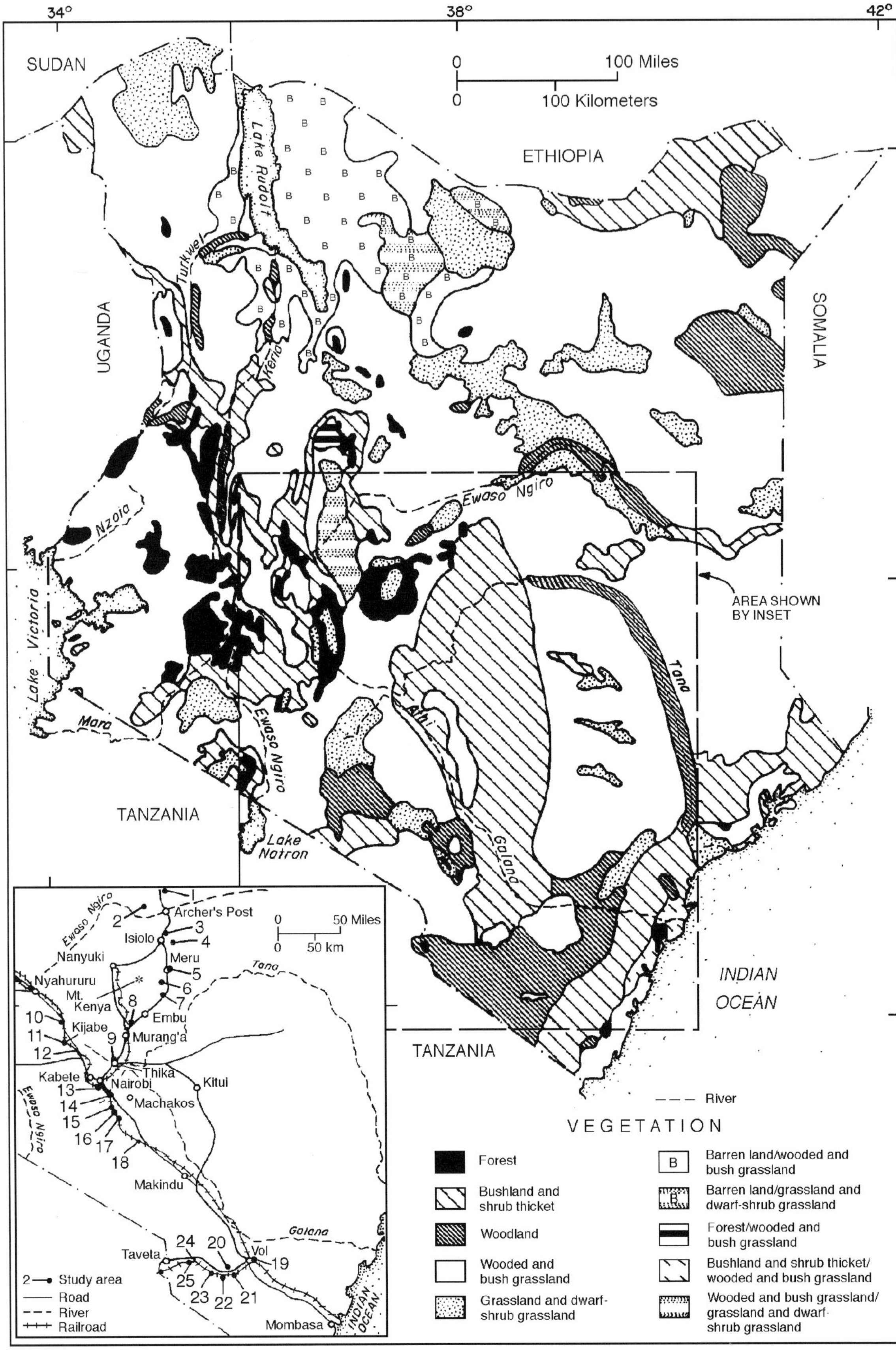

Frontispiece. Vegetation map of Kenya (After Pratt and Gwynne 1977). Inset shows study areas in which 49 sets of matched photographs are located.

Kenya's Changing Landscape

Raymond M. Turner, H. Awala Ochung', and Jeanne B. Turner

The University of Arizona Press

Tucson

Published from camera-ready copy prepared under the authors' supervision by

Rose Houk, manuscript editor
CAD Unlimited / Charles Sternberg, cartographer
Donna Opocensky, designer and compositor

The University of Arizona Press

∞ This book is printed on acid-free, archival-quality paper.
Manufactured in the United States of America

03 02 01 00 99 98 6 5 4 3 2 1

Library of Congress Cataloging-in-Publication Data
Turner, R. M. (Raymond M.)
Kenya's changing landscape / Raymond M. Turner, H. Awala Ochung',
and Jeanne B. Turner.
p. cm.
Includes bibliographical references (p.) and index.
ISBN 0-8165-1871-8 (pbk.: acid-free paper)
1. Vegetation dynamics—Kenya—History—Pictorial works. 2. Plant ecology—Kenya—History—Pictorial works. 3. Vegetation and climate—Kenya—History—Pictorial. 4. Landscape changes—Kenya—History—Pictorial works. I. Ochung', H. II. Turner, Jeanne B.
III. Title.
QK402.K4T87 1998
581.7'096762—dc21 97-45290
CIP

British Library Cataloging-in-Publication Data
A catalogue record for this book is available from the British Library.

Foreword

For a large part of its history, ecology has sought to understand and characterize systems "in equilibrium." This is a necessary and understandable first step in the study of complex systems. However, the enduring difficulty of this deceptively simple conceptual task has revealed that equilibrium over any appreciable period of time is extremely rare and that it is perhaps more a useful construct than an actual observable state.

Together with the realization that natural systems tend to be inherently variable has come a growing awareness that some of the observed changes in natural systems may in fact be directional and may be driven by more profound changes in climate that might be the consequence of human agency. Thus, to add to the challenge of understanding climate-driven biological processes, long-term changes in global climate are probably overlaid on other patterns of short-term climate variability.

Determining whether or not there has been change and, if so, what the direction and magnitude of change might be is a task that is far from trivial. First there are the noted patterns within patterns of climate variability – some cyclic, some apparently chaotic, and some directional, all often operating at different temporal and spatial scales. This makes it difficult to determine where we might be at any one point in time. A second, and perhaps more fundamental, challenge to the study of environmental change is the general lack of a comprehensive environmental record by which to capture change. Long-term systematic observations of vegetation are exceedingly rare. Instrumental records of climate only became generally available during this century, and coverage for many areas remains spotty. Thus, a major challenge to the study of environmental change has been to develop reliable ways of reconstructing what existed in the past from the little evidence that has come down to us.

Linkages between climate and vegetation are so basic that terms used to describe them are often used almost interchangeably (e.g., rainforest; steppe; desert). Thus, one method used to reconstruct past climates is the study of past vegetation patterns as proxies of climate. The evidence used may be parts of the plants themselves, captured and preserved in any number of ways that are seldom obvious (e.g., protected rodent nests or the pollen contained in lake sediments). Another type of evidence that is both more obvious and more compelling has come available over the past century and one-half and which serves as a central feature of Kenya's Changing Landscape: the use of historical photographs.

Photographs, in many ways, speak volumes about past conditions. They often allow definitive statements to be made about vegetation type, individual plant species, their size and abundance, as well as their distribution across the landscape. It is also possible to infer the types of disturbance to which the site might been subject, through other clues contained in the photograph – the presence of dwellings and agricultural fields, or the stone monuments that mark a mining claim, suggest specific types of disturbance. In a more subtle way, it is also possible to develop a less-well-defined sense of the landscape and how it is organized that cannot be derived from a written description, no matter how eloquent. This is the power of photography.

The study of recent vegetation change – such as revealed through historical photography – introduces another confounding factor: the influence of human activity. Beginning from the basic relationship between climate and vegetation, it may be assumed that changes in vegetation must indicate a change in climate. However, it is obvious that human activities can also affect vegetation either directly (e.g., agricultural clearing) or indirectly (e.g., fire generation/suppression). Thus, in addition to the cyclic, chaotic and directional influences of climate on vegetation, there is often unpredictable and unreported human activities that help shape vegetation patterns.

As a consequence of these factors, it must be realized that there is also a great deal that pictures cannot convey. How representative is the site of conditions that existed in the region? What is the trend in vegetation condition at the time of the photograph? What did the site look like 20 years earlier? What uses has this land been subjected to within the past decade or the past century? These are the issues which dictate that historical photography can only be used with great care – a care that is easily overlooked but well demonstrated in this volume.

Kenya's Changing Landscape is a landmark work for several reasons. First, it represents a significant historical event in that it is the combined work of the two scientists (Homer Shantz and Raymond Turner) most responsible for developing and applying matched photography to the study of environmental change. Second, beyond its methodological significance, it is important in what it tells us about the nature of vegetation change. By often using three points in time, this work allows for a more confident assessment of the often non-linear trajectory of vegetation change over a period that was marked by great ecological, economic and demographic change. Moreover, its historical depth allows us to appreciate how quickly and how profoundly environments change. Finally, it is significant because it captures change in all its diversity. Kenya's Changing Landscape provides a poignant view of both the causes and consequences of environmental change but also suggests what these changes might mean to local populations. Like few volumes that have preceded it, Kenya's Changing Landscape captures the essence of environmental change and very often puts a human face to it.

Charles F. Hutchinson
Tucson, Arizona

Acknowledgments

Special appreciation is due Theuri J. Njoka, Range Management Department, University of Nairobi, without whose interest and support this work could not have been done. Several employees of the Kenya Wildlife Service supported, in various ways, our activities within Kenya, its national parks and preserves. Among these are Dadson Mugwe, George Tokro and James Else. We thank Geoffrey Mungai of the East African Herbarium, Nairobi, for assistance with plant identification and in answering taxonomic questions.

To Charles T. Mason, Curator of the Herbarium (Retired), University of Arizona, we owe a large debt of gratitude for providing initial access to the Shantz photographic collection, and to Lucinda A. McDade and Philip D. Jenkins for later continuation of this courtesy.

Michael D. Gwynne, United Nations Environment Programme (Nairobi); M. John Makin, Natural Resources Institute (UK); and Reginald Rose-Innes, Overseas Development Program, (retired) (UK) provided intellectual backing during the early planning and later field stages, as did Garry F. Rogers, who was active in planning this program although unable to participate in later field activities. We thank Patrick Kamonyi for his assistance during our field excursions.

Thanks are given to reviewers Julio L. Betancourt, Janice E. Bowers, Timm Hoffman, Paul S. Martin, Garry F. Rogers, Billie Lee Turner I, and Robert H. Webb for their many helpful comments. In addition, Webb and Betancourt responded to our call for help in seeing this book through the publication process in a manner far beyond the call of duty. To Kennith E. Foster and Charles F. Hutchinson, Director and Associate Director, respectively, Office of Arid Lands Studies, University of Arizona, we express thanks for the interest and generous support they gave at the crucial publication stage of this project.

We are grateful to G. Robert Lofgren (Laboratory of Tree Ring Research, University of Arizona) and Semon Gachuiri (Kenya Department of Meteorology) for providing the Kenyan climate data. The first draft of the map was computer generated at CAD unlimited (Tucson, Arizona) with the generous support of Justin M. Turner of Arroyo Engineering (Tucson, Arizona). Charles Sternberg drafted the revised map. Donna Opocensky's careful attention to detail and computer skills are acknowledged with appreciation. Peter Griffiths and Tade Orr provided essential computer support during numerous stages of this desktop publishing adventure. Rose Houk labored over a near final manuscript, proving abundantly that authorship may not embrace all matters of consistency and style. This rephotography survey project was made possible by a grant from the National Geographic Society, which we gratefully acknowledge.

CONTENTS

APPENDIXES

PLATES

Introduction

Kenya is a country of extreme landscapes, a diverse biota, and a varied people. Until a century ago it was part of the "undeveloped, unsettled frontier" as viewed by 19th-century Europeans. Such continents as Africa and North America had been important objects of the expansionist policies of European nations as they divided the terrain of faraway, lightly inhabited and lightly developed lands. As the colonial powers gained control, they brought about dramatic changes in the landscape. In Kenya these changes, some of extraordinary proportions, have been telescoped into the span of the last century.

Many observers have described the highlights of this recent history. As an addition to these chronicles, we present here a look back through the last three-quarters of this tumultuous period, using views recorded through the camera lens at the beginning, middle, and end of that critical slice of time. The repeat photography method, unexcelled for assessing changes in open landscapes such as savanna, grassland, and desert, provides the basis for contrasting changes in Kenya with those we have seen in North America. Striking differences exist on the two continents: in the population histories of wild and domestic herbivores, in patterns of human occupation, and in land management practices. The replicated photographs we present provide an opportunity to contrast the effects these forces may have had on the current landscape status of the two different areas.

The use of paired or matched photographs, viewing a given scene at two or more points in time, has become a widely used technique for examining landscape change (Rogers et al. 1984). Some of the earliest landscape photographs were taken during the United States Civil War. Shortly after the end of that conflict, several of the war's photographers travelled through the American West as members of railroad survey parties. By the late 1860's, they had recorded many western landscape views on photographic film. During the 1870's, natural resource survey parties of the newly opened West usually included photographers. By the turn of the century, landscape photography had come into its own and thousands of scenes, depicting conditions of the time, had been captured on film. As time passed, interest in reexamining past conditions increased.

Ironically the first extensive account of landscape change based upon matched photographs was not from North America, but from Africa. The book, *Photographic Documentation of Vegetational Change in Africa over a Third of a Century* (Shantz and Turner 1958), was the first of its kind (Rogers et al. 1984).

The circumstances which led to an African first in the new field of repeat photography were due to Homer L. Shantz, renowned botanist, and former president of the University of Arizona (1929-1936) and the Ecological Society of America (1928). Born in Michigan in 1876, Shantz attended high school and college in Colorado and received his doctorate at the University of Nebraska in 1904. He took many photographs of eastern Colorado as early as 1904 and 1905 (McGinnies et al. 1991).

Shantz made two trips to Africa in the early 1920's, traversing the entire continent both times. On these trips he took nearly 4500 photographs, many of which illustrate the natural plant cover. With this archive base, he returned to Africa in 1956-1957 with B. L. Turner, to relocate earlier camera stations, to repeat the photographs, and to describe any landscape changes that had occurred during the intervening one-third of a century. In all, 241 photographs were successfully matched between Cape Town, South Africa, and the countries of Kenya, Uganda, and Belgian Congo, now known as the Democratic Republic of the Congo (Shantz and Turner 1958). In part because of civil unrest, Ethiopia, Sudan, and Egypt were excluded.

The present study provides a third look at 17 of the sites previously visited by Shantz and a second look at 32. Most of the latter were originally visited in the 1920's. During the 72 years since Shantz' initial African journey, many political and demographic changes have taken place on the continent, culminating in new country names, new boundaries, and greatly expanded rural and urban populations. Although we initially contemplated duplicating Shantz' entire Cape Town-to-Cairo route, we settled upon Kenya, a country where the success of a resurvey could be assured. Many hundreds of Shantz' photographs covering other areas in Africa await use by other workers. They repose at the Shantz Photographic Archive in the Herbarium at the University of Arizona in Tucson.

Shantz 1920 trip

Shantz' visit to Kenya in 1920 was sponsored by the Smithsonian Institution in conjunction with the Universal Film Company. Shantz worked for the Office of Foreign Seed and Plant Introduction, Bureau of Plant Industry, U. S. Department of Agriculture. The party left Brooklyn, New York, on July 16, 1919. With Shantz were Edmond Heller (expedition director), H. C. Raven (Smithsonian field naturalist), and six others with roles listed as business manager, actor and director, News Enterprise Association member, camera man, developer, and photographic equipment supervisor. The team crossed the African continent from south to north, from Cape Town to Cairo, via a route along or near the eastern seaboard. The party was in Kenya from April 19 to July 20, 1920.

Shantz carried two cameras: a Century Grand Senior 4 x 5 and a Goerz Kodak 3 1/4 x 4 1/4. He took three lenses for the Century Grand Senior: a Goerz Dagor f6.8, a wide angle lens, and a Goerz telephoto lens; and for the Goerz Kodak, a Celor f4.8 (Shantz n.d.). He took 2650 photographs in Africa, of which 456 were taken in Kenya. Perhaps one-third of his landscape images are suitable for photographic matching.

Shantz 1924 trip

The 1924 expedition was an outgrowth of the first but under auspices of The Educational Commission to East Africa sponsored by the Phelps Stokes Fund, the British Colonial Office, and the Church Missionary Society. The aim was to take still and moving pictures and for Shantz to continue his study of African countries, their people, agriculture, soils, fauna, and flora (Shantz n. d.). The route included the African countries known today as Somalia, Ethiopia, Djibouti, Tanzania, Kenya, Uganda, the Democratic Republic of the Congo, Malawi, Zambia, Mozambique, Zimbabwe, Botswana, and South Africa. Shantz shot roughly 900 still photographs on this trip, of which 213 were taken in Kenya during February and March.

Shantz-Turner 1956-1957 trip

This trip was sponsored by the United States Office of Naval Research and administered through the College of Agriculture, University of Arizona. Shantz was accompanied by Billie Lee Turner (not related to the senior author), botanist from the University of Texas, Austin. They left Cape Town, South Africa, in September 1956, following much of Shantz' 1920 route northward into East Africa. By March 1957 they had reached Uganda. They took 524 photographs; 63 were in Kenya. Of the roughly 150 early Kenyan photographs that were suitable, they matched 37 and published 13 (Shantz and Turner 1958).

Trip of 1992

Our trip was sponsored by the National Geographic Society and was undertaken in collaboration with the Range Management Department, University of Nairobi. We succeeded in matching 45 photographs that were first taken in 1920 or 1924. Eleven of these are among the 13 published following the 1957 Shantz-Turner visit to Kenya (Shantz and Turner 1958). In addition, we matched 4 photographs first taken in 1957, for a total of 49 photographic matches. We decided against travelling in two areas visited by Shantz where in 1992 security was a problem. These are the regions between Nakuru and Kisumu in western Kenya and the area beyond about kilometer 15 north of Archer's Post in northern Kenya. Several photographs in these areas await matching during calmer times.

The Setting

Kenya is located in the east-central part of the African continent. Straddling the equator, the country extends from roughly 4° north latitude to 4° south. The area of Kenya is 582,646 square kilometers with elevations ranging from sea level to 5,190 meters at the summit of Mount Kenya.

Agriculture is the principal economic base of the country, with coffee and tea the primary cash crops. The percentage of the land under cultivation is relatively small; 87 percent is open rangeland (Pratt and Gwynne 1977). Rainfed cropping is feasible only in the humid west, the central highlands, and the narrow coastal strip.

The People

The condition of Kenya's landscape today is heavily influenced by the country's political and economic history, especially since about 1890 when Great Britain assumed control. Traditional African societies relied upon rudimentary technology to raise agricultural goods mostly for their own needs. They had no formal monetary system, made largely irrelevant because of the lack of external trade. Children were educated and socialized without formal schooling which their elders viewed as unessential in their pastoral economy. In time, the Europeans preempted much of the more productive agricultural and grazing land and in many ways altered the cultural ecology of the native peoples. Such colonial pressures have contributed to the degradation of resources in some instances (Sindiga 1984, Tignor 1976). In other instances, improvement has resulted (Tiffen et al. 1994).

We rephotographed lands occupied primarily by three aboriginal cultures: the Maasai/Samburu, Kamba, and Kikuyu. Differences in these cultures and the potential of the lands they occupy have dictated the effects of political vicissitudes during the 72 years spanned by the photographs. Much has been written about the history of these land transfers and their impact on the native peoples, primarily the Maasai (Glover and Gwynne 1961, Ward and White 1971, Talbot 1972, Tignor 1976, Evangelou 1984, Collett 1987, Lindsay 1987, Sindiga 1984, Homewood and Rodgers 1991, Western 1994), but also the Kamba (Owako 1971, Ward and White 1971, Bernard 1972, Tignor 1976, Tiffen et al. 1994), and the Kikuyu (Ward and White 1971, Tignor 1976).

As an unfortunate prelude to the post-1900 influx of European settler-farmers, the people of Kenya experienced harsh times during the 1890's. Possibly the most devastating occurrence, especially for the Maasai, was the outbreak of rinderpest, a disease of cattle and wild bovids that entered the African continent in 1889 and crossed Kenya in 1890. As the result of this epidemic, most Maasai cattle died. This disaster was joined by a smallpox epidemic, a severe drought, and a famine. The human population declined abruptly.

The European encroachment at the beginning of the 20th century followed at the end of this difficult period, just as human, livestock, and wildlife numbers were recovering from the series of disasters. Much land may have appeared unoccupied. At the time of their maximum influence in the 1800's, the Maasai population was estimated at about 45,000. (These numbers include those living in present-day Tanzania.) By the mid-1920's, having recovered from the disasters, the population was roughly 50,000; and by 1961, it had reached 117,000 (Talbot 1972). Consequently, by the time of Shantz' first visit in 1920, the Maasai had recovered from their hardships of the 1890's, but many hectares of their previous holdings had already been relinquished and the newly alienated land was being turned to uses quite different from those of earlier times.

The Maasai were affected by these losses since seasonal grazing of certain elevated grassland areas was no longer possible under the new ownership. Restriction of grazing to smaller, more arid areas resulted in loss of vegetation and soil erosion (Glover and Gwynne 1961). Other pressures stemming from the influx of Europeans resulted in considerable Maasai forest land becoming part of the shifting cultivation system of Maasai neighbors, further shrinking lands available for livestock grazing.

Camera stations which view Maasai land are those located from Isiolo north to Archer's Post (this land is occupied by the Samburu, a group related to the Maasai), the station near Taveta, and westerly views from the railroad tracks south of Nairobi that separate Maasai lands from Kamba lands (Odingo 1971). The Maasai life-style has been one of seminomadism and transhumance based upon the requirements and movements of livestock, with modification when defense or warfare dictated. Apart from livestock predators, these people do not normally kill wild animals, nor do they eat meat, obtaining dietary protein from their livestock in the form of blood and milk. Cattle are mostly not raised for meat or for sale, and as animals accrue, they add to the owners' wealth, prestige, and security.

Although the Maasai lost more land to the European settlers than did other native people, the Kikuyu and Kamba also suffered. The Kikuyu occupied the highlands around the west base of Mount Kenya southward to Kiambu district just north of Nairobi. By 1934, with land alienation largely completed, 325 square kilometers of Kikuyu land had been encroached upon by

Europeans (Tignor 1976). Our camera stations from Thika north to Meru and from Kijabe northwesterly to Longonot lie within the area dominated by the Kikuyu.

The Kamba lands pertinent to this report lie southeast of Nairobi. This upland area, with Machakos as its population center, had been occupied by these people for perhaps two centuries before the arrival of the Europeans. The Kamba lived mainly by raising livestock, especially cattle, and by practicing shifting agriculture in the production of grain and pulses. They traded actively with others, cattle being the mainstay of their economy (Tiffen et al. 1994). The Nairobi-Mombasa highway and the Kenya Railroad right-of-way mark the southeastern edge of these lands. Our photograph stations from Athi Plains to Sultan Hamud traverse a line separating Kamba lands toward the northeast from those of their neighbors, the Maasai, toward the southwest.

The 45-year period from 1915 to 1960, roughly corresponding to the period between Shantz' 1920 and 1957 visits, saw the number of Europeans increase from about 5,500 to 61,000 (Sorrenson 1968, Wasserman 1976). Only a few thousand of these actually owned agricultural lands, but the influx serves to illustrate the rate at which Kenya attracted white settlers.

Following Kenya's attainment of independence in 1963, much of the preempted "scheduled" land of the highlands was, over time, bought out by the government, returned to the Africans, and divided into small subsistence farms. Land redistribution is still in progress and is a long-standing major political issue (Wasserman 1976, McCarthy and Mwangi 1982). Thus, before we took the latest series of photographs, a few years to several decades may have elapsed since return of the lands to the Africans.

At 3.7 percent per year, the growth rate of Kenya's population is one of the fastest in the world. Its 1992 population of 26.2 million is expected to reach 44.8 million by 2010 and 62.3 million by 2025 (Anonymous 1992). What effect these human population changes will have upon the Kenyan landscape is at the heart of this study. The photographs span a period that began when there were roughly 30 persons per square kilometer; by the end of the period, 300 persons per square kilometer were settled on the land. During the same period, transportation was greatly improved and the nation's economy had strengthened through less dependence on subsistence farming and a greater involvement in foreign trade. The photographs reveal changes during this period and may help anticipate what the country will be like in 2025, for example, when the projected population density will reach 717 persons per square kilometer.

Climate

Most of Kenya is dry. Roughly 25 percent of the country receives less than 250 mm of precipitation each year; more than 70 percent receives less than 560 mm. Although the central highlands offer a cool, bracing climate, these almost ideal conditions for human occupancy prevail over only a small fraction of the country's total area. Less than 15 percent of the Kenyan landscape receives more than 760 mm of rain. Annual rainfall is erratic, especially in the arid areas where variations are most extreme.

For its relatively small size, Kenya has a surprising variety of rainfall regimes. Near the equator, two wet seasons are the rule, with a tendency as the Tropics of Cancer and Capricorn are approached for a single summer maximum to prevail. This generality is upset along the equator, however, where the east Kenyan Highlands have bimodal rains, and the west Highlands a single-season regime. The difference in these two rainfall patterns has a marked effect upon crop and livestock production. Areas with two growing seasons promoted by sparse bimodal rainfall will experience elevated livestock production compared to areas with the same amount of rainfall occurring in only one season. In contrast, crop production would be possible only under the single season regime and would fail where annual rainfall values of 500 mm or less are distributed over two distinct seasons (Pratt and Gwynne 1977, LeHouérou 1989).

Soils

The stations shown in our photographs are mostly found on soils derived from Tertiary to Recent volcanics or on those derived from Precambrian and Cambrian rocks. Tertiary ash deposits and lava flows, produced as the Kenyan Highlands were being formed by upwarping of plateau surfaces, add complexity to this interior elevated region (Saggerson 1969). The tectonic disturbances that have profoundly affected the relief of Kenya are those responsible for the great Rift Valley and for block faulting such as that forming the Taita Hills. The Precambrian and Cambrian basement complex rocks, occurring in a north-south trending band between 90 and 150 km wide, lie to the east and north of the highlands. Many of these rocks are high in quartz, producing coarse, well-drained soils with relatively low fertility. In contrast, broad depressions and level plains with a mantle of colluvium may develop soil with impeded drainage. An extreme example of this type of soil is "black cotton" soil which, due to its high content of montmorillonite clay, cracks deeply when dry and swells when wet. The upper horizon is black as a result of the high content of organic matter (Scott 1969).

Vegetation

Several vegetation maps and vegetation classification schemes have been proposed for East Africa and are reviewed by Pratt and Gwynne (1977). To group our photographs in a logical sequence, we use the vegetation classification scheme they propose (frontispiece). The vegetation of Kenya coincides broadly with the country's annual rainfall pattern. Forests occur within a narrow coastal band of high rainfall and within other areas of high rainfall in the central highlands and in the highlands bordering Uganda. The rest of the country is mostly arid or semiarid and supports vegetation referred to as woodland, brushland, and grassland.

The photographs are presented generally from arid to moist conditions. They are grouped according to vegetation units, a brief description of which appears at the beginning of each grouping. In each instance, the status of the vegetation in the early photograph determines its placement. We assumed that the early photographs usually represented conditions most free of human impacts.

Description of Commonly Encountered Plants

Of the several hundred plant species encountered in this study (Appendix 3), a few are notable because of their frequent occurrence, abundance, or change at camera stations. We list below the 26 species most frequently encountered at the camera stations. These plants have been observed at six or more of the sites. The ecological significance of each is discussed in further detail below, and the number of sites at which the plants are located is shown in parentheses. Authors' names for these species are given in Appendix 3.

All 12 grasses are native, as is the forb; both shrubs are introduced; and of 11 trees, two (Tasmanian blue gum, silky oak) are introduced species.

Grasses

Kenya needle grass *(Aristida keniensis).* An increaser or invader in denuded or heavily grazed areas (Kemei 1986, Pratt and Gwynne 1977), this species was usually found at heavily grazed or disturbed situations, where annual precipitation ranges from 400 to 600 mm. It is palatable only when young. (6)

Sweet pitted grass *(Bothriochloa insculpta).* The plant is a tall perennial with a strong tendency to spread by stolons when grazed or cut or when colonizing bare ground (Pratt and Gwynne 1977). It has a wide ecological amplitude, including black-cotton soils. (8)

African fox tail *(Cenchrus ciliaris).* Known as buffel grass in some countries, this drought-tolerant, fire resistant perennial grass, of medium height and often with woody stems, occupies areas with 500 mm to 600 mm rainfall. (10)

Common star grass *(Cynodon dactylon).* Widely distributed in Kenya (Edwards and Bogden 1951), this low, sward-forming grass produces both rhizomes and stolons, features that have made it an aggressive weed in warmer parts of the Americas. Known as Bermuda grass in many parts of the world, it stands up well under heavy grazing pressure and is often associated with disturbed areas. (10)

Naivasha star grass *(Cynodon plectostachyum).* A spreading perennial grass with stout, rapidly growing stolons, it is found at elevations between 600 and 2000 m, particularly in the *Acacia-Themeda* scattered-tree grassland. It also occurs in open areas of the high bush and in semidesert grasslands, where it is found chiefly in favorable situations such as the beds of intermittent streams. Its grazing value is rated as good (Edwards and Bogdan 1951). (10)

Hook grass *(Latipes senegalensis).* A prolific seed producer of only medium grazing value (Bogdan 1958), this grass occurs in grasslands and bushlands, on eroded slopes, and in rocky places. Its productivity declines following burning (Thomas and Pratt 1967). (6)

Guinea grass *(Panicum maximum).* Widely distributed in Kenya from sea level to 2000 m, guinea grass is seldom found in the open and typically forms colonies beneath or near shrubs and trees. Productivity of this grass increases following burning (Thomas and Pratt 1967). Before becoming too coarse at maturity, it is considered a valuable forage plant for grazing or for cutting as green fodder (Edwards and Bogdan 1951). (15)

Kikuyu grass *(Pennisetum clandestinum).* Commonly occurring in moister regions above about 2000 m, this short, leafy, and nutritious species flourishes under intensive grazing. It forms dense sod, spreading by both rhizomes and stolons (Pratt and Gwynne 1977). (9)

Bamboo grass *(Pennisetum mezianum).* A coarse woody plant with wiry stems when mature, bamboo grass is palatable when young. While often the dominant in open or wooded grasslands on vertisols from 800 to 2000 m elevation (Pratt and Gwynne 1977), it also occurs at the moist fringe of the semidesert grasslands (Edwards and Bogdan 1951). (10)

Elephant grass *(Pennisetum purpureum).* Producing bamboo-like clumps and attaining heights of 3 m to 5 m, the plant has stems up to 2.5 cm in diameter with leaves 3.5 cm wide and more than 1 m long. It occurs in moist regions from sea level to 1800 m where it is often found along river banks and at forest edges. Widely used as a reserve green fodder and as silage when cut before the stems become hard and woody, it is capable of very high yields under conditions of good fertility and proper management (Rose-Innes 1977). Elephant grass does not produce viable seeds and is established through vegetative propagation from stem cuttings (Edwards and Bogdan 1951). It is widely planted to check erosion and for soil buildup. (7)

Maasai grass *(Pennisetum stramineum).* A perennial that forms loose tufts, this grass spreads by means of stout rhizomes. It is found at elevations between 1000 and 3000 m in bushed grassland. It usually grows on black-cotton soils where it normally forms large colonies. It is palatable and has good seeding qualities (Edwards and Bogdan 1951). (11)

Red oat grass *(Themeda triandra).* This perennial is probably the most widespread grass in East Africa. Its grains germinates after fires (Belsky 1992, Pratt and Gwynne 1977, Lock and Milburn 1977, Lind and Morrison 1974, Trollope 1982). The awns on the grains are hygroscopic and undergo twisting movements with changes in humidity. This motion buries the grain in the soil to depths where it is protected from the effect of fires. Germination is apparently induced by the increased light and soil temperature of newly burned areas. Regularly found on black-clay soils, this plant is a persistent constituent of edaphic grasslands. On other soils, the density of this grass declines in just a few years if fires are not repeated. This grass does not withstand heavy grazing (Heady 1966) and disappears when grazing increases just to the point where fires are prevented (Lind and Morrison 1974). The grass occurs in most ecological zones except the semidesert grassland. It is absent above 3000 m and is of little importance below 1200 m (Edwards and Bogdan 1951). (8)

Forb

Sodom apple *(Solanum incanum).* A large, ruderal forb, attaining a height of 1.5 m, this plant is widely distributed in Kenya, especially in bushed grassland and rarely in semidesert grasslands. Unpalatable to livestock, it is common in abandoned fields, along roadsides, and in heavily grazed or poorly managed pastures (Edwards and Bogdan 1951). (8)

Shrubs

Curse of India *(Lantana camara).* An herbaceous plant that may become a shrub, it attains heights of 1-3 m or more. Introduced from tropical North America, perhaps by way of India, this aggressive noxious plant has become naturalized and widespread in Kenya, often growing in disturbed ground next to crop land. It can be distinguished from the closely related indigenous species by its prickly stems and showier flowers which may be pink, yellow, orange or white (Dale and Greenway 1961). (10)

Mexican sunflower *(Tithonia diversifolia).* A native of Central America and Mexico, this unpalatable shrub has become widespread in Kenya. A member of the sunflower family, attaining heights up to 3 m, it has large, deep-yellow flower heads. It commonly grows along roadsides and field margins. As a measure of the rate of its spread, B. L. Turner (personal communication 1994) comments that during the 1956-1957 trans-African trip, he does not recall seeing this plant in any African country nor does it appear in his field notes, despite his great interest at the time in members of this family. (10)

Trees

Whistling thorn *(Acacia drepanolobium).* A dwarf to medium-sized tree, attaining heights of 1.5 to 7.5 m. It has straight thorns, some of which become basally inflated, hollow, and are colonized by ants (*Crematogaster* sp.). These specialized thorns produce a whistling noise in strong winds as the air passes the entry holes to the formicary. When disturbed by browsing animals, the ants rush from their homes in the thorns and may discourage even large herbivores such as giraffes from taking more than a few bites from the host plant (Dagg and Foster 1976). This plant is highly favored by giraffes, and browsing may be so heavy from above by these tall animals that the plant height is kept low, forming an inverse browse line (Dagg and Foster 1976). The shrub is sensitive to fire and its presence in large numbers in a grassland is an indication of infrequent burning. The plant is common on seasonally waterlogged black-cotton soils at elevations ranging from 600 m to 2550 m. (6)

White thorn *(Acacia hockii).* A frequent occupant of hilly terrain, this small tree is widespread at intermediate elevations ranging from 750 m to 2250 m. Seldom forming thickets, it is usually found as scattered individuals among other trees and shrubs and often encroaches upon overgrazed grassland (Coe and Beentje 1991) but may be kept in check by frequent burning (Dale and Greenway 1961). It has bright yellow flowers and dark rough bark that peels off in papery sheets. (6)

Egyptian thorn *(Acacia nilotica).* Usually a small tree with a dense, mushroom-shaped crown, it reduces range productivity by casting such heavy shade that shade-intolerant grasses do not grow well beneath it (Bogdan and Pratt 1961). Under favorable growing conditions this tree may reach a height of 20 m. Elephants are fond of this tree, lifting the bark with their tusks. Goats eat the flowers (Baumer 1983). Chiefly inhabiting bushed grassland, it occurs along the coast of Kenya and at low to medium elevations (1000 m to more than 2000 m). The fruits are indehiscent and germination of the enclosed seeds is apparently enhanced by passage through the gut of large herbivores (Pratt 1967). When eaten in large quantities, the seeds cause nerve disorders among herbivores (Bogdan and Pratt 1961). It is little affected by fire (Pratt and Gwynne 1977). (7)

Umbrella thorn *(Acacia tortilis).* A medium to large, flat-topped tree with dense crown, this is very common in semiarid areas below elevations of about 1500 m. Very resistant to drought, this tree often grows where the water table is deep; but because of its great adaptability, it may also occur in alluvial valleys or along permanent streams where it may reach a height of 20 m (Baumer 1983). Sensitive to fire, this tree only regenerates in areas where fire does not occur for five consecutive years (Lamprey 1979). The pods, which are spirally twisted or semicircular in shape, are a valuable source of crude protein to herbivores. Because lower branches soon fall away, the crown does not interfere with growth of shade-intolerant forage plants beneath. The foliage is a valuable browse for livestock. This and other *Acacia* species may grow as much as one meter a year and then become tolerant of fire after a few years. Where giraffes are numerous, their browsing may keep pace with the trees' top growth, maintaining the plants' vulnerability to fires for extended periods (White 1983). Impalas are partial to the indehiscent fleshy fruits (Pratt and Gwynne 1977), and germination of the enclosed seeds is enhanced by passage through the gut of large herbivores (Pratt 1967). (16)

Desert date *(Balanites aegyptiaca).* Common on black-cotton soils, this small, dense-crowned tree attains heights of 4.5 m to 6 m, or exceptionally to 10.5 m. Its spherical crown is a dense mass of long thorny twigs. These twigs and the dense evergreen foliage provide valuable browse for many herbivores (Dale and Greenway 1961), and are especially relished by camels. The fruits are sought after by elephants and baboons (Baumer 1983). In keeping with its evergreen habit, its roots are deep-lying with little lateral spread (Pratt and Gwynne 1977). (12)

Bridelia *(Bridelia micrantha).* A tree to 12 m high with spiny branches and dense wide-spreading crown. Its wood is valued as fuel and for carpentry. It is often found at forest edges or in riparian habitats (Dale and Greenway 1961). (6)

Broad-leaved croton *(Croton macrostachyus).* Although attaining heights of 24 m, this tree is usually 12 m tall or less. It is common in moist habitats at elevations between 600 m and 2000 m (Dale and Greenway 1961). (6)

Red-hot poker tree *(Erythrina abyssinica).* A deciduous tree to 12 m tall. Its bark has blunt woody spines, and its stout branches bear strong recurved thorns. This tree is common in areas with at least moderate rainfall (Dale and Greenway 1961). (9)

Tasmanian blue gum *(Eucalyptus globulus).* This fast-growing Australian native, attaining heights of 100 m in Africa, occurs in plantation forests, parks, and roadsides throughout all but the driest parts of Kenya. The bark peels off in large strips and may become a fire hazard where not removed from the understory (Schmutterer 1976). (7)

Bark cloth fig *(Ficus natalensis).* The Kikuyu revere this tree as the abode of ancestral spirits (Dale and Greenway 1961). Free-growing or epiphytic on other trees, it may attain heights of 12 m. Its bark is gray and smooth, its foliage dense and evergreen. Aerial roots descend from major branches. (6)

Silky oak *(Grevillea robusta).* A robust Australian native, this tree grows to heights of 30 m. It has been introduced throughout the East African highlands as a shade tree for coffee and tea plantations, and in parks, roadsides, and gardens (Schmutterer 1976). (15)

Methods

The black-and-white photographs from 1992 were taken with Kodak T-Max 100 or T-Max 400 120 roll film using a Graflex XL camera, usually fitted with an 80 mm Zeiss Planar lens, although a few photographs were taken with a Zeiss Tessar 100 mm lens. The black-and-white film was developed commercially in Nairobi. Photographic prints were made by R. Turner in a home darkroom upon return to Arizona. An essential part of our equipment was a Polaroid film back with which we obtained proof prints in the field. On these we labelled the more conspicuous plants and terrain features. As a backup camera, we used a 35 mm Canon AE1 with a Promaster 28-105 zoom lens and Kodak T-Max 100 and 400 film. Most scenes were duplicated in color using Kodachrome 64 film in a Rollei 35 mm camera with 40 mm lens. These negatives and diapositives are stored at the U.S. Geological Survey Research Projects Office, the Desert Laboratory, Tucson, Arizona. The Shantz negatives are filed at the Herbarium, University of Arizona, Tucson, Arizona.

Shantz left little information to locate the camera stations, often labelling the negatives imprecisely as, for example, "between Meru and Ewaso Ng'iro" or "this side of Meru." In most instances this was not a problem because his itinerary was recorded in field notes, and the photographs usually show the distinctive profile of local hills or mountains that are readily visible from the roads travelled. In a few instances where the roads have been moved or where Shantz left the main track to take photographs, relocation of the camera stations was difficult or, in some instances, not feasible. All of our camera stations are shown on 1:50,000-scale or, in a few cases, 1:250,000-scale maps filed with the photographs. Global coordinates for the camera stations are given in Appendix 1. Detailed information regarding the photographs is provided in Appendix 2. The repeat photography field and darkroom techniques that we used are described in Rogers et al. (1984).

Field activities usually took the following form. Once the camera was in position to match as closely as possible the old scene, we exposed a Polaroid print to verify the camera position and test our estimate of proper film exposure. With camera in place, we exposed images with regular black-and-white and color film; the Polaroid print was labelled; we noted current features, changes detected, comments of local inhabitants, and other relevant items, in our field book. Description of the scene usually followed the sequence foreground, midground, background to ensure complete and thorough coverage. Plant identification, Polaroid print labelling, and field note writing was largely the responsibility of Ochung'; J. Turner photographed the scenes on 35 mm color film and kept detailed photographic notes for all film exposed; R. Turner jockeyed the camera tripod into final position and exposed the black-and-white film, after which the entire team walked about the area sharing and recording thoughts about the landscape changes.

Although requiring several abrupt jumps from one part of the country to another, we sequence our study sites according to the type of vegetation in which they occur. The reason for this presentation order is to group together sites representing roughly similar habitats having the potential for comparable changes. The usual order is in some instances interrupted by the effect of extremes in topography where, for example, moist forests might occur along valleys crossing unforested dry uplands. Here the regional low rainfall is ignored, and the topographically induced forested sites are grouped with those falling under higher rainfall regimes. As an aid to the reader, we have assigned to the camera stations a "study area" number corresponding to a restricted area of no more than a few square kilometers where one or more photographs were taken. Thus, the 49 photographic views (plates) are grouped into 25 study areas, with as many as six separate views in a given area. These numbers, assigned sequentially from north to south, simplify locating the plate numbers.

A comprehensive checklist of the major plant species was compiled at each photograph site. This inventory (Appendix 3) will provide a useful benchmark for evaluating future change at these same sites. Many plant species shown in the photographs are often not identified, and minor species may not appear in the views. The value of the photographic record can be greatly enhanced when this additional information is provided.

To characterize the rainfall at the camera stations, we selected long-term rainfall data for the nearest stations. For the transect south and east of Nairobi, we have chosen Voi, Machakos, and Makindu; for the region near Nairobi, two short overlapping records from the station at Kabete and at Kenyatta Airport provide values for the years 1930-1963 and 1951-1992, respectively; and for the segment north of Nairobi, stations at Murang'a (Fort Hall), Embu, Meru, Isiolo, and Archer's Post are used (See Appendix 6).

To judge the degree to which climate is responsible for the status of vegetation at the time of the photographs, we use annual departures from the long-term average rainfall and normalized rainfall indices for the two major rainy seasons to estimate the seriousness of drought. Long rains are those occurring March through May; short rains occur October through December. The rainfall index (RI) is defined as:

$$RI = \frac{P - \overline{X}}{S}$$

where P = seasonal precipitation, $\overline{X}$ the long-term average for that season, and S the standard deviation from P. A value of RI=1.0 is used to denote severe drought. These values are given for 10 stations located in the region of the camera stations (Appendix 6).

By using the rainfall indices, moisture conditions for any period, such as the years immediately preceding the photographs, can be readily evaluated. The use of a rainfall index in which the values are expressed as standard deviation units from the long-term average serves to deemphasize absolute values and to simplify comparisons between stations having vastly different rainfall amounts and degrees of rainfall variability.

DWARF-SHRUB GRASSLAND

Included here are lands where the typical sparse grassland is beset with low shrubs not exceeding 70 cm in height. Widely scattered larger shrubs or stunted trees may be present. Fires occur rarely (Pratt and Gwynne 1977). Our four stations in the dwarf-shrub grassland lie in a region having soils derived from the Cambrian "basement" rock (Scott 1969). Upland areas are coarse, well drained, and relatively infertile. Soils of the low-lying bottomlands are either saline or alkaline and support specialized vegetation.

A.

B.

C.

Ol Doinyo Sabachi (Plate1, Study Area 1)

A, (1920). The camera station is 8.5 km north-northwest of present-day Archer's Post along route E821 to Kauro and Laisamis. The view is across the Larisoro Valley toward Mount Lololokwi (Ol Doinyo Sabachi). The dominant taller shrub is Volkensinia prostrata. The smaller shrub with more numerous stems is Indigofera spinosa, as is the small heavily browsed plant dotting the openings. These two species are common occupants of alkaline soils (Pratt and Gwynne 1977). Perennialgrasses, perhaps dropseed and/or hook grass, commingle with the same shrubs as in the foreground setting. This is a dwarf-shrub grassland without any overstory of large shrubs or trees. Umbrella thorn is the dominant tree lining the distant water course. (Shantz and Turner 1958, Fig. 68a.)

B, (1957). The termite hill at left has changed form, having lost the sharp spires and becoming rounded. Although difficult to see in this view, the second termite hill at right is still present. *Indigofera spinosa*, a plant that increases with heavy grazing, is dominant now. *Volkensinia prostrata* is still present but is of less importance than in 1920 (Shantz and Turner 1958). The perennial grasses present 37 years earlier are almost completely gone, except for a few individuals of dropseed, three-awn, and hook grass still growing in the vicinity. A single small plant of umbrella thorn can be seen at right. The background is unchanged. (Shantz and Turner 1958, Fig. 68b).

C, (1992). The two termite hills are much smaller than before. The umbrella thorn tree to the right of the closer termite mound probably persists from the earlier visit. Most of the plants now visible are *Indigofera spinosa* with lesser numbers of *Sericocomopsis hildebrandtii* and *Saddera hirsuta. Volkensinia prostrata* is absent from the area. The only grasses present are the perennial *Oropetium capense* and the annual Kenya needle grass, which are little used by large herbivores. Umbrella thorn has increased in importance. Both umbrella thorn and three-thorned acacia occur along the dry river bed in the background.

Discussion. At the time of the earliest view, perennial grasses were conspicuous and the dwarf shrubland of the fore- and midground was free of small trees. In 1957, after 37 years, the grass cover was almost completely lost and spiny bushes and small trees had increased (Shantz and Turner 1958). This continued to 1992, as the dwarf-shrub grassland of 1920 was being converted to wooded bushland. In addition to the changes induced by herbivores, the area is used by the military whose off-road vehicle tracks are numerous.

A.

B.

C.

Nakadeli (Plate 2, Study Area 2)

A, (1920). This and the two following camera stations are located within a few hundred meters of each other on the Samburu National Reserve, 1.5 km south-southwest of the Archer's Post gate. In this view Nakadeli Hill is at the right. The conspicuous foreground plant with sword-like leaves is Ehrenberg wild sisal. The shrubs in the immediate foreground are *Volkensinia prostrata.* At midground, *Volkensinia* and *Ipomoea cicatricosa* are the dominant plants with a scattering of a wiry grass that is probably hook grass. A few small plants of umbrella thorn are scattered across the scene in the background (Shantz and Turner 1958, Fig. 65a.) Note the termite nest at extreme right.

B, (1957). The camera is turned slightly too far right to capture all of the previous scene and may actually be a few meters from the original position. Ehrenberg wild sisal is present in the foreground at the extreme right. The dark plant in the foreground is *Ipomoea donaldsonii.* Farther back the conspicuous spiny shrub is *Commiphora candidula.* The dominant small shrubs are *Ipomoea cicatricosa* and *Volkensinia prostrata.* The herbaceous plants of the spaces between shrubs are the forb *Blepharis linariifolia* and annual and short-lived perennial grasses such as six-weeks three-awn, stinkgrass, and hook grass. Umbrella thorn has conspicuously increased on the level area in back and is accompanied by an occasional false umbrella thorn. The site adjoins a frequently used ford of the Ewaso Ng'iro and shows heavy grazing by livestock (Shantz and Turner 1958, Fig. 65b).

C, (1992). The dominant small shrubs are now *Sericocomopsis hildebrandtii* and *Duosperma eremophilum.* The small leafless shrub in right foreground is *Ipomoea* sp. Grasses present include hook grass and Kenya needle grass. Umbrella thorn is the dominant small tree and is joined by lesser numbers of false umbrella thorn. The *Commiphora* seen in the 1957 view is gone. The termite hill, first noted in 1920, is no longer present. Many bare patches of soil and numerous game trails occur.

Discussion. The site lies within the Samburu National Reserve, established in 1962 (Wilson 1989). Formerly rangeland receiving heavy use by domestic and wild herbivores, the area now is largely closed to domestic livestock and probably receives heavy use by native herbivores. The wild animals are managed to promote tourism, and the large number of animals results in trampling, heavy grazing, and browsing. The site's position next to a major river assures its heavy use by all classes of native herbivores including elephant, zebra, buffalo, and giraffe, and the smaller bushbuck, impala, kongoni, and dik-dik. Large woody plants are increasing and the perennial grasses are gone, barring a few "stumps" of *Enteropogon macrostachyus* hidden among acacia bushes. *Volkensinia prostrata,* the former small shrub dominant, is now absent. The herbaceous dominants of the area are now annual grasses of little forage value.

A.

B.

C.

Nakadeli Ranger Post (Plate 3, Study Area 2)

A, (1920). The prominent hill at center is Nakadeli. The most important plant of the foreground is *Salsola dendroides,* associated with saline soils. The low shrubs in the open space at right are probably *Volkensinia prostrata*. Note the open grassland strip at the base of the hillslope. A trail crosses the scene in the foreground. Heavily grazed tufts of the perennial hook grass are evident (Shantz and Turner 1958, Fig. 67a).

B, (1957). The camera station has moved forward perhaps as much as 35 m. The vegetation of the saline flat is much as before but the formerly grassy swath at the base of the hill is now dominated by small, unidentified trees (Shantz and Turner 1958, Fig. 67b).

C, (1992). Atop the small knoll on the right are buildings used as a ranger's post by George Adamson when Samburu National Reserve was established. The buildings are now abandoned, and the gate to the reserve they overlooked is now located 1.5 km to the northeast. The camera station is approximately the same as for the 1920 view. The most conspicuous change in the saline flat of the foreground is the increase in *Salsola dendroides*. While acacias have increased in adjacent areas, they are still lacking here. Grasses include feather fingergrass, hook grass, and stinkgrass. Both *Aloe dawei* and *Euphorbia heterochroma* are restricted to sites beneath shrub canopies. The tree, first seen in 1957 as a new occupant of the narrow band of grassland seen in 1920, is false umbrella thorn. A mix of umbrella thorn and false umbrella thorn occupies the steeper slopes.

Discussion. The area appears heavily utilized. Perennial grasses occur only as low tufts *(Chrysopogon aucheri)* or as waifs *(Enteropogon macrostachyus and Chrysopogon aucheri)* hidden under the canopy of shrubs. A sparse cover of forbs and weak annual grasses occupies openings. The site's proximity to the river is responsible for its heavy use by the diverse herbivorous fauna of the area. The loose and friable topsoil is readily moved by the wind. The shrubs *Blepharis linariifolia* and *Salsola dendroides* are increasing; *Volkensinia* is now rare at the site. Beyond the saline flat, false umbrella thorn and umbrella thorn have continued to increase with time. Although relatively unpalatable to cattle, false umbrella thorn is used somewhat by camels. In certain areas it is heavily cut for firewood (Lusigi 1984).

A.

B.

C.

Old Archer's Post (Plate 4, Study Area 2)

A, (1920). Located in the immediate vicinity of Plate 2, the view from this station is southeasterly overlooking the hill named Marti Girgir. The buildings on the hilltop are part of the original Archer's Post. The road crossing the view at the base of the hill leads to a nearby ford of the Ewaso Ng'iro. Note the termite nest at far right midground. The low shrub between the camera and the hill base is mainly *Volkensinia prostrata*. A small plant of *Commiphora candidula* is at left midground. The leafless light-barked shrub in right center is probably *Ipomoea cicatricosa*. Scattered shrubs on the hill and at its base are mostly umbrella thorn. The grass forming the small clumps between the *Volkensinia* plants is probably hook grass (Shantz and Turner 1958, Fig. 66a).

B, (1957). Archer's Post has been abandoned. The ground between the camera and the hillslope is still dominated by *Volkensinia prostrata* and *Ipomoea cicatricosa. Ipomoea donaldsonii* appears to have increased in importance. The conspicuous small tree, umbrella thorn, has increased significantly since 1920, changing the aspect from shrubland to wooded shrubland. False umbrella thorn is a minor element here. The *Commiphora* at left persists and appears to give protective cover to a cactus like plant, *Caralluma retrospeciens*. At far left the tufted perennial, hook grass, is still present but appears to be exceeded in importance by annual grasses such as six-weeks three-awn, stinkgrass, and *Chloris pilosa* (Shantz and Turner 1958). The apparent increase in grass cover is probably due to the different seasons the two views represent (Shantz and Turner 1958, Fig. 66b). The termite nest is no longer visible.

C, (1992). The road present in 1920 is gone. The camera is within a few meters of the original position. *Sericocomopsis hildebrandtii* and *Duosperma eremophilum* are the small shrub dominants. Umbrella thorn has greatly increased in spite of uprooting, browsing, and breakage by elephants. False umbrella thorn is scattered throughout this developing bushland. Animal trails form a conspicuous network across the foreground.

Discussion. *Volkensinia* is now absent from this view, although a few individuals still occur within a few meters of the camera station. *Volkensinia* and *Ipomoea cicatricosa* have given way to *Sericocomopsis hildebrandtii* and *Duosperma eremophilum*, the current small shrub dominants. The open wooded bushland of umbrella thorn, at an early stage of development in 1920, has increasingly dominated the midground. This increase has occurred in spite of uprooting by elephants and heavy pruning by a variety of large herbivores. The game trails are more conspicuous in 1992 than in the earlier views. The observed decline of *Volkensinia* is consistent with observations at three nearby sites, indicating that it is disappearing from a large area.

GRASSLAND

Grasslands are dominated by grasses and perhaps other herbs. Shrubs and trees, which may be present as widely scattered individuals or groups, contribute no more than 2 percent of the total ground cover. Periodic fires are usual. Where coverage of a scattering of individual or of grouped trees is no more than 20 percent, grasslands are known as wooded grassland. Periodic fires are the usual condition. Another grassland variant, bush(ed) grassland, is characterized by having various proportions of both scattered trees and shrubs with a combined canopy coverage of less than 20 percent. Periodic fires are the rule (Pratt and Gwynne 1977). Many grassland sites are seasonally flooded or have impeded drainage. When heavily charged with clay, such flooded soils are shunned by large herbivores because they may become trapped in the deep mire.

Konza Station (Plate 5, Camera Station 17)

A, (1920). Photograph taken at Konza Station on the Kenya Railroad, 75 km southeast of Nairobi. This view, labelled simply "Detail of Themeda grassland" (Shantz n.d., p. 169), shows an essentially featureless grass-covered plain with an occasional desert date in the distance. The camera was probably on a tripod set on the ground, accounting for the lower vantage point than that of most other photographs taken on the railroad. The grasses are Maasai grass mixed with red oat grass.

B, (1992). Except for a few isolated plants of Maasai grass and red oat grass, the original grassland near the camera station has been replaced by a cultivated field. The original camera position is behind one of the buildings at Konza Station; the new position is within a few meters of the original. The foreground grasses are common star grass and Kikuyu grass. Crops grown here include corn, beans, and potatoes. Beyond the field is a grassland of red oat grass, perhaps little changed from earlier times apart from an increase of whistling thorn and desert date. Toward the background hill are found white thorn and *Combretum* sp. Additional plants not noted above are silky oak, pepper tree, Egyptian thorn, crab grass, Kenya needle grass, and feather fingergrass.

Discussion. This site bears the usual stamp of settled areas: plantings of exotic trees with herbaceous ruderal species appearing in heavily disturbed areas. The uncultivated area beyond the corn field remains grassland although with an increased woody component.

Kaputei Plains (Plate 6, Study Area 16)

A, (1920). View over the Kaputei Plains from a point 18 km northwest of Konza Station along the Kenya Railway right-of-way. This elevated view, taken from a moving railroad car, shows individuals of whistling thorn scattered over a grassy plain.

B, (1992). The camera, positioned on a tripod in the center of the railroad tracks, is 2 m to 3 m lower than was Shantz' camera. Even with the different perspective, close comparison with the earlier view is possible. The right-of-way is now fenced and supports a dense growth of common star grass, Guinea grass, sweet pitted grass, and red oat grass. Woody plants visible across the fence are mainly whistling thorn and desert date, the dark rounded tree.

Discussion. In 1992 prevalent old culms of perennial grasses indicate the area seems not to have been burned or impacted by large herbivores. Across the fence an increase in both grass and woody vegetation is evident. Whistling thorn and desert date are both more abundant. The grass is tall enough to hide the base of many shrubs.

Ziwani Siding (Plate 7, Study Area 25)

A, (1920). This station is located at Ziwani Siding, almost exactly on the west boundary of Tsavo West National Park and 14 km by road east of Taveta. The photograph, taken from the railroad right-of-way, shows Salaita Hill. Judging from the photograph's clarity, the train had stopped at the siding when Shantz took this picture. This area was the scene of intense fighting between British and German forces during World War I, and a rock wall erected as fortification can be seen circling the hill just above midslope. The foreground shows a recently cultivated field of intercropped beans and corn. The field gives way at midground to an open bushed grassland. The tree at left is sandpaper cordia. Other trees visible are not easily identifiable from the photographs but include umbrella thorn, desert date, and three-thorned acacia. The grass cover is probably mostly African fox tail, Naivasha star grass, and red oat grass.

B, (1992). The soil of the entire fore- and midground area is a claypan which represents the exposed B horizon of the original soil profile. The foreground is no longer cultivated, perhaps because of the loss of topsoil. The vegetation is now bushland. Many of the trees present earlier, including the cordia, have been cut down. The foreground shrubs are mainly *Hoslundia opposita* and *Hermania alhiensis* with fewer *Grewia tenax.* The shrub at far left (crown extending above the skyline) is desert date. The heavily pruned shrub to the right of the utility pole is *Cordia gharaf;* that to the left is *Boscia salicifolia.* The grass cover, although sparse, is rich in species including African fox tail, hook grass, stinkgrass, and Naivasha star grass. All of the perennial grasses are confined to the cover of shrubs and trees. The openings between the woody plants are partly covered by annual grasses such as stinkgrass. The grass cover of Salaita Hill is mainly African fox tail, hairy herringbone grass, and *Tetrapogon tenellus.* The scattered trees and shrubs on the hill, now more numerous, are mainly three-thorned acacia and *Grewia bicolor.*

Discussion. The site is within open range heavily utilized by domestic livestock of the Maasai. Indications of heavy use include spaces between shrubs that are either bare or support annual grasses of low forage value, confinement of more valuable perennial grasses to protected sites, loss of topsoil, and signs of excessive trampling by cattle. The bushed grassland of 1920 has been converted to bushland where woody plants have replaced grasses as the dominants and where the potential for fires is greatly reduced.

Plain of Ngare Mara I (Plate 8, Study Area 3)

A, (1920) View across the Ngare Mara Plain toward the Mula and Nyambeni Hills from beside the main road 17 km north of present-day Isiolo. The dark band of trees at left background line Ngare Mara, an intermittently flowing tributary of the Ewaso Ng'iro. Shantz thought the soil of this level plain perhaps had a "lime" underlayer. He described the vegetation as short grass and noted oryx, cheetah, and zebra near the camera station (Shantz n.d., field notes, entry for June 15, p. 20). The foreground grass is mostly bamboo grass mixed with Maasai grass. The cattle belong to the Samburu, the region's dominant tribe. Note the two small leafless shrubs in the right foreground.

B, (1992). Exact location of the original camera station was made difficult because dense, tall shrubs and trees block the view of the distant mountains. The new camera station is within 100 m of the original and now occupies an area completely transformed from a grassland to a wooded bushland where trees, shrubs, and herbaceous plants comprise 15 percent, 50 percent, and 10 percent, respectively, of the ground cover, with the remaining 25 percent bare soil. The tall, flat-crowned trees are umbrella thorn; although tall, none appear old. Other trees and shrubs are false umbrella thorn (four tall shrubs at center midground and one at right foreground), *Acacia nubica,* desert date, and *Grewia tenax.* Perennial grasses, found mostly beneath the protective cover of acacias and desert date, include the following: African fox tail, Maasai grass, and bamboo grass. The low white shrubs in the foreground are *Sericocomopsis hildebrandtii.*

Discussion. In 1920, the grass appears heavily used even in this June scene at the end of the long rains. Excessive trampling by domestic and wild herbivores has exposed large patches of soil, some 2 m to 3 m square. The foreground shrubs may represent an incipient stage of ensuing changes. By 1992 the grass cover is gone; the dominants are woody trees and shrubs while forbs of low forage value, such as *Barleria acanthoides* and the short, spiky *Solanum dubium*, appear to have increased.

Plain of Ngare Mara II (Plate 9, Study Area 3)

A, (1920). This station is about 65 m northwest of the previous station and is within 100 m of Shantz' original position. The view is a 90° turn to the left from the previous and looks across the Ngare Mara Plain and the distant Ngare Mara River toward the Shaba Hills in the middle background. The trees bordering the Ngare Mara are visible. Between that water course and the camera station is an extensive level plain of heavy clay soil dominated by bamboo grass, which appears heavily grazed.

B, (1992). As with the other photographs taken in the vicinity, positioning the camera at this station was difficult because of the rank growth of tall plants blocking views of the distant mountains. Accordingly, the camera station in 1992 was adjusted so that a small part of the mountainous skyline can be seen. The clump of shrubs at right is actually a mixture of *Sericocomopsis hildebrandtii* and *Cadaba farinosa.* In the foreground are scattered, closely cropped ruderal grasses and the spiny forb *Barleria acanthoides.* All perennial grasses are mostly confined to the protective cover of the acacias and desert dates. Such grasses include African fox tail, Maasai grass, *Chrysopogon aucheri,* and bamboo grass. Species that have increased presumably as a result of heavy use are the forbs *Solanum dubium* and *Barleria acanthoides,* the small shrub *Sericocomopsis hildebrandtii,* and the annual grasses hairy herringbone grass and *Aristida mutabilis.* Patches of open bare soil are more conspicuous now than in 1920.

Discussion. Significant change has occurred. The former grassland is now represented by scattered remnants growing within the protective cover of woody plants that are new since 1920. Grazers will do less well now than browsers. The local Samburu probably graze fewer cattle and more goats than in 1920. Trampling is more apparent in 1992 than earlier, possibly because of heavier use by goats. The shape of some trees and shrubs reflects browsing by Megaherbivores such as elephants, giraffes, and camels, yet pressure from these animals has not inhibited encroachment of the woody plants.

Plain of Ngare Mara III (Plate 10, Study Area 3)

A, (1920). This station is a few hundred meters farther north than the previous two and closer to the Ngare Mara. In the background can be seen the Mula Hills in front of the taller, more distant Nyambeni Hills. The vegetation here is a bushed grassland. *Acacia nubica* appears in the foreground and as scattered individuals farther back. The large trees with dense crowns are desert date. The grass is Maasai grass, characteristic of sites with impeded drainage and seasonal flooding.

B, (1992). Position not exact but within 100 m of original. The photograph is taken from the elevated roadbed, built with soil removed from the foreground borrow pit. The excavation has promoted considerable erosion. The vegetation, a bushland thicket, is dominated by false umbrella thorn with lesser amounts of *Acacia nubica*. The tall, flat-topped trees are umbrella thorns. *Acacia zanzibarica* occurs along the disturbed roadside. Other plants seen here are listed in the description for the first Ngare Mara station.

Discussion. The woody component of the site beyond the borrow pit has increased and the grass cover decreased. Exact comparisons are not possible because the 1992 view does not closely match the earlier one, but the perennial grass cover has probably been significantly altered with the dense, almost continuous cover of Maasai grass in 1920 giving way to widely scattered clumps of perennial grasses confined wholly to protected sites. These grasses include Maasai grass, bamboo grass, and *Chrysopogon aucheri*. The vegetation has changed from bushed grassland to bushland thicket.

Stony Athi Station (Plate 11, Study Area 15)

A, (1920). The camera station is along the railroad right-of-way, 4.2 km southeast of Stony Athi Station and 3 m northwest of kilometer marker 474.3. The Mua Hills are in the left background. The foreground is grassy and without shrubs; the midground vegetation contains a scattering of whistling thorn but can still be regarded as grassland.

B, (1992). Taken from a tripod on the roadbed, this view provides a poorer perspective than the original where the camera was 2 m to 3 m higher. The right-of-way, now fenced, is dominated by red oat grass with lesser amounts of bamboo grass. The spindly, light-stemmed shrub at left foreground is *Hibiscus aponeurus.* Several individuals of whistling thorn are visible in the fenced right-or-way. *Aspilia pluriseta,* a small shrub mostly hidden by the grass cover, grows abundantly here. Beyond the fence the composition is much the same. The two dark, rounded shrubs are desert date. The vegetation physiognomy is bushed grassland.

Discussion. The site seems heavily affected by large herbivores. Desert date, a low tree much used by giraffes, is neatly pruned. The stunted habit of the whistling thorn further indicates heavy herbivore pressure on the area. The grass cover appears denser and shrubs more abundant in 1992 than 72 years earlier. *Aspilia pluriseta,* now a subdominant of the area, shows no loss of foliage to megaherbivores and may be increasing as a result. The apparent significance of fire in shaping this grassland was noted by comparing the vegetation of this site with that directly behind the camera. The barren railroad right-of-way apparently serves as an effective fire break resulting in two neighboring tracts with presumably different fire histories. The site behind the camera and across the railroad track lacks desert date, red oat grass is denser, whistling thorn is less common, and *Aspilia* nearly absent. We speculate that this site has been burned more recently or frequently than the one pictured here. Grants gazelle and ostrich were seen in the vicinity.

Athi Plains (Plate 12, Study Area 13)

A, (1920). This photograph was taken along the railroad about 6 km northwest of the Athi River bridge, near Athi River Station. The view is across the Athi Plains toward the Athi River, marked by the dark band of trees. The vegetation is grassland with widely scattered shrubs. The shrub on the foreground hill is whistling thorn. This plant is also seen at center beyond the grassy expanse from the hill. The dense shrubs beyond the whistling thorn may be desert date. Wires strung from poles along the railroad right-of-way appear at the top.

B, (1992). This photograph is probably taken within a few meters of the original position, except that the camera is now positioned lower than before. Beyond the fence, and extending as far as the Athi River, is Nairobi National Park, established in 1946. The foreground shrubs are mainly *Grewia similis,* although an Egyptian thorn plant can be seen just left of center and one of *Maytenus putterlickioides* at right of center. The foreground grass is *Cymbopogon validus.* Additional grasses found in the right-of-way are red oat grass, tanglehead, Guinea grass, bamboo grass, Maasai grass, African fox tail, and sweet pitted grass. The two dark shrubs just beyond the fence at left center are desert date. Most of the other shrubs between the fence and the river are whistling thorn, although scattered individuals of Egyptian thorn, desert date, and several other species occur. The dominant grasses of the plain, which is now regarded as a bushed grassland, are red oat grass and *Hyparrhenia collina.* The trees along the banks of the distant Athi River, fever tree and white thorn, appear larger now. Beyond the river are fields of wheat and corn interspersed with homesteads. The overhead wires are gone except for a single sagging one.

Discussion. Increased density of shrubs, both in the right-of-way and beyond, is the most conspicuous difference between the picture pair. The plain has changed from grassland, with scattered woody plants, to a bushed grassland with a much greater complement of these long-lived species. No shrubs from the early view seem to persist to 1992, suggesting that some force such as fire or browsing animals has removed them, although the net change over 72 years has been toward increased woody plants in the assemblage. The century-long absence of elephants in Nairobi National Park may be responsible for the increased size of trees along the Athi River, an effect elephant removal elsewhere has had on riverine communities (Owen-Smith 1988). The park is the dry-season home for many large herbivorous mammals, many of which disperse into the nearby Athi and Kaputei Plains during the remainder of the year (Deshmukh 1986). Cattle ranches and fencing adjacent to the park diminish the area available to the native animals during the wet season (Talbot 1972).

Sultan Hamud (Plate 13, Study Area 18)

A, (1920). This site is 2.6 km south of Sultan Hamud along the Mombasa-Nairobi Railroad. The camera position is on the roadbed between the tracks 5 m northeast of railroad marker 397.3. Kiou Hill is at the left and Kilungu/Nunguni Hills appear in right background. This photograph was taken from a moving train (Shantz n. d.). The grass cover is short and somewhat sparse. The population of the dominant shrub, whistling thorn, is represented by individuals of all size classes, indicating that recruitment of new individuals into the population has occurred steadily for many years. The shape and height of these plants (the plant at right foreground is probably about 3 m tall) indicate that pressure from giraffes and other browsers is slight. The grass is bamboo grass.

B, (1992). The foreground is now planted to corn, beans, sorghum, and pigeon peas. Beyond the field and the paved highway, the vegetation is composed of several species of acacia, especially whistling thorn, as well as desert date. This vegetation extends across the midground to the background slopes where it gives way to a different bushland dominated by Natal rhus, velvet-leaved combretum, and umbrella thorn. Additional plants on the black-cotton plain are Egyptian thorn, white thorn, three-thorned acacia, hook thorn, bamboo grass, Guinea grass, and African fox tail.

Discussion. The multiaged population of whistling thorn in the first view suggests that it had been fire-free for many years despite its proximity to the railroad and the constant threat of fire started by sparks from the wood-burning locomotives of the day. Perhaps the grass cover had been too light to carry fire. The vegetation beyond the cultivated ground and paved highway is taller in 1992 than before. Some patches of cultivated ground are visible on the background hills, but poor print quality does not allow their status to be determined for 1920. The site trend of the midground is toward bushland in 1992 from bushed grassland earlier. The ground cover today comprises numerous perennial grasses, forbs, and shrubs growing beneath the whistling thorn. The site appears to receive only light use by large herbivores.

Longonot (Plate 14, Study Area 10)

A, (1920). The camera station is within 100 m of the railroad station at Longonot and appears to be within the railroad right-of-way. The bare area in front of the camera station is probably the borrow pit dug during construction of the railroad, which was completed in 1900 (Miller 1987). Plants just beyond the low vertical bank are the perennial grasses, bamboo grass, and Naivasha star grass; the forb, *Aspilia mossambicensis;* and the shrub, whistling thorn. The midground area is a grassland of bamboo grass sparsely dotted by low plants of whistling thorn. The hill and the low ridges flanking it are covered by bushed grassland with a few taller trees visible along the ridgetop.

B, (1992). The camera, now located on top of a berm forward of its original position, looks down upon a field of wheat. At left is a field of corn; across the remainder of the view is more wheat. Among the trees just beyond the fields are introduced pepper tree, Tasmanian blue gum, silky oak, and the indigenous fever tree. Numerous houses are now evident. The hill and ridges have widely scattered dwellings with adjoining plots of wheat, corn, beans, and potatoes. Uncultivated areas appear heavily grazed by domestic stock. Steeper parts of the hill support patches of leleshwa. Steep-walled erosion gullies are present on the slope to the right of the hill.

Discussion. Human impact is conspicuous throughout, grading from total alteration of the original plant cover in the foreground to partial alteration in the distance. Soil tilling and crop production on more level terrain give way to less intense use such as livestock grazing and fuel-wood gathering on the steep slopes.

Mbuyuni (Plate 15, Study Area 24)

A, (1920). This camera station, lying between railroad kilometer markers 93.8 and 93.9, is at the railroad workers' camp called Mbuyuni, 10.3 km east of Ziwani Siding. The view is northerly across the Serengeti Plain. Shantz noted when he took this photograph, ". . . out across the plain below station. Grass, acacia and baobab" (Shantz n.d., p. 159). Eight km farther he notes, "Great quantity of lion tracks along track" (loc. cit.). The dense grass cover of this bushed grassland is dominated by Guinea grass and Naivasha star grass. In addition to the baobab noted by Shantz, the tall trees are probably three-thorned acacia and umbrella thorn.

B, (1992). Within a few meters of the original position, the camera station now lies within the Tsavo West National Park; domestic livestock are excluded. Located away from a water source that might attract the wild herbivores of the area, the site is lightly grazed. The dense cover of grass is dominated by Naivasha star grass with lesser amounts of Kenya needle grass, African fox tail, hook grass, and stinkgrass. Trees visible in the background include umbrella thorn, three-thorned acacia, and hook thorn. The tall tree at extreme left background is three-thorned acacia. The vegetation physiognomy is bush grassland.

Discussion. Although domestic livestock are no longer permitted on the area, wood cutting is apparently not restricted. Tall trees are fewer in 1992 than before, having been cut, perhaps, for charcoal production. Many of the trees are still present but in shorter coppiced form. Browsing by wild herbivores is obvious, especially on acacia and desert date. The grass cover appears little changed over 72 years.

Isiolo-Meru I (Plate 16, Study Area 4)

A, (1920). This and the following three camera stations are at a remote site visited by Shantz to view a rhinoceros killed by a hunter. Shantz noted ". . . a great deal of game . . . " including "Oryx, ostrich, rhino, gazelle and zebra." Ol Doinyo Degishu Range is seen in the distance. The town of Isiolo lies on the distant plain at extreme right. Vegetation in the vicinity of the camera station is dominated by Maasai grass, red oat grass, and sweet pitted grass, and by an open stand of umbrella thorn. The light-colored, spiny, leafless shrub at left foreground is *Maytenus putterlickioides*. This tract with numerous volcanic rocks has not been cultivated although the level ground beyond the rim of the midground hill shows some signs of clearing.

B, (1992). The camera location is probably within a few meters of the original. Umbrella thorn is still the dominant large plant although many of the original trees have been cut and are represented by dense clumps of basal sprouts. Other woody plants of the area include desert date, common rhus, and melia. Forbs of the area include *Barleria acanthoides,* Sodom apple and caltrops. Grasses, although abundant, are heavily utilized. These include Maasai grass, bamboo grass, Naivasha star grass, and red oat grass. The immediate foreground has been mostly cleared of rocks and, although cultivated at one time, now lies fallow.

Discussion. See Discussion for Plate 19.

Isiolo-Meru II (Plate 17, Study Area 4)

A, (1920). In this view, which looks to the right roughly 80° from near the same point as the previous photograph set, Mount Shaba appears left of center in the background. The soils are derived from recent volcanics. The people at extreme left, one standing and the other kneeling or sitting, appear to be examining the rhinoceros that was killed here. Umbrella thorn is more abundant, but mostly of smaller stature, than in the previous view. The small shrubs have probably become established within the last 20 to 30 years. In the foreground are Maasai grass, bamboo grass, and sweet pitted grass.

B, (1992). The foreground area is an abandoned field and now supports a scattering of Maasai grass and *Chloris pycnothrix*. Shrubs of the area include *Grewia similis* and desert date.

Discussion. See Discussion for Plate 19.

Isiolo-Meru IIIa (Plate 18, Study Area 4)

A, (1920). The photographs for Plates 18 and 19 were taken from the same point which is just a few meters from the position for the previous two. The view is southwest across rolling hills with Loloria Hill at left. Photographed with a long focal-length lens, this site was used by Shantz as an example of acacia-tall grass savanna (Shantz 1940, Fig. 113). The dense grass cover is composed primarily of Maasai grass and *Hyparrhenia hirta*. The woody element is predominantly umbrella thorn. Scattered throughout is a shrub, partially concealed by the grass cover, which may be small umbrella thorn. The two trees closest to the camera station at right show evidence of having survived a fire, not surprising given that much of this vegetation is burned during the annual drought period (Shantz 1940). Note the summit of Mount Kenya showing above the cloud bank at left.

B, (1992). The foreground is an abandoned field in which the low, mat-forming caltrops is the dominant plant. The grass clumps are of Maasai grass and bamboo grass. Most of the acacias have been cleared. The vegetation has lost its orchardlike aspect and shows a strong component of shrubs as it changes from bushed grassland to dwarf-shrub grassland. Mount Kenya is covered by clouds.

Discussion. See discussion for Plate 19..

Isiolo-Meru IIIb (Plate 19, Study Area 4)

A, (1920). This view, taken with a normal lens and looking slightly to right of the previous, overlaps it by about 75 percent. The dark discoloration at the trunk bases, noted on these same trees in the previous view, is seen here as well. Shantz does not mention fire in his field notes but does refer to the wind-trained trees.

B, (1992). Note the houses at right midground. Other features as noted for Plate 18.

Discussion. Numerous homesteads, cultivated fields, and pastures dot the area in 1992. Most big game has probably left. Fields have been cleared of rocks and some woody growth. The dominant tree, umbrella thorn, has been heavily used to construct brush fences around cultivated plots and livestock corrals. Local inhabitants relate that during periods of drought, the acacia trees are cut for charcoal production as a cash crop to replace the failed cereal crops that would normally provide some income. Accelerated soil erosion is apparent, especially where cultivation has largely depleted the topsoil, exposing the lower horizons.

Sagana (Plate 20, Study Area 8)

A, (1920). The view is southeasterly toward Kiambicho Hill from a point that in 1992 is in the center of the village of Sagana. The earlier photograph was taken in a dense grassland extending from the foreground to the tree-lined Tana River in the background. The grasses were Guinea grass, pyramid dropseed, and jungle rice. The foreground and midground are moist areas. The mounds are termite nests.

B, (1992). Buildings now occupy the original camera station; consequently, it was moved a few meters to the right. The Nairobi to Nanyuki railroad, appearing at right, was constructed in 1927. The town of Sagana was established on the railroad at an important road crossing. The low dense grass along both sides of the road is common star grass. Curse of India is the dark shrub at right, between the dirt road and the railroad, and at left midground. Trees in the Kiambicho Forest on the background mountain include red-hot poker tree and white thorn. In the cultivated area between the village and the hill are fields of beans, cassava, corn, mango, and banana.

Discussion. Evidence of urbanization dominates the camera station. Several species of eucalyptus are the dominant trees in the village. Between the village and Kiambicho Hill numerous estates are interspersed among cultivated fields, with grazing by domestic livestock on the hill itself.

Kimathi (Plate 21, Study Area 8)

A, (1920). This site is west of Sagana near Kimathi. The background slopes are covered by the Kiamuti Forest. The grass in the foreground is Guinea grass. Grassland extends from the camera across the valley to hills beyond. The scattered trees cannot be identified, although the one at right, which persists to the present, is red-hot poker tree. The valley floor is mostly uncultivated grassland with a few cultivated areas at right midground. Rooftops of a village are visible at the foot of the slope left of center.

B, (1992). This photograph is taken from the roadside where the weed *Triumfetta rhomboidea* is seen at left foreground. A small plant of finger euphorbia is in mid-foreground. The field of corn intercropped with beans in the foreground is but one of many that are new across the valley since 1920. Mangoes, bananas, and pigeon peas are crops of secondary importance. Besides the houses at midground, numerous settlements dot the hillslope in the background. The large red-hot poker tree in the 1920 view has been trimmed back to make space for a building but can still be seen above the taller part of the structure at right of center. The tall trees at midground are mainly silky oak. The shrubs *Leucas calostachyus* and curse of India are common along the roadside at left and the fence rows. The poorly drained areas of the valley floor are dominated by adrue. The background hills are covered by a mixture of native red-hot poker tree, introduced silky oak, and lesser amounts of native white thorn. Grasses include Guinea grass and common star grass.

Discussion. Although some features are only dimly seen in the original photograph, the changes over 72 years are nonetheless starkly revealed when the two views are compared. Settlements are now common throughout areas bordering the valley floor. The valley is more heavily cropped than before with fields of beans and corn. An introduced tree, silky oak, has been sparsely planted throughout the cultivated area. At the base of the far hills, the density of three-thorned acacia, Egyptian thorn, and white thorn appears to have increased.

Meru (Plate 22, Study Area 5)

A, (1920). This station is located about 1 km northwest of the center of Meru. The camera station overlooks a tributary of the Kathita River. The distant mountain at left, Mbokoro Peak, is surrounded by the Thuuri Forest. The area is grassland broken by linear parklands of large trees. The grass is Kikuyu grass and the trees are mostly bark cloth fig mixed with large-leaved cordia. The white flowered plant in the foreground is *Cleome* sp. (Shantz n.d., p. 198). The large-leaved shrub at scattered places on the foreground ridge is Sodom apple. Although some cultivation is apparent on the slope at right, the rolling terrain is unfenced rangeland. Houses are visible at several places across the view at midground.

B, (1992). Camera position is not exact. From the true location, roughly 50 m upslope to the left, the view of the distant mountains is completely blocked by tree plantings. The present position, although inexact, does provide a limited view of the background. Haze and clouds were so dense that even with a yellow filter the film could not record the background mountains. The most obvious plants are mainly introductions from other continents. The trunk at right is silky oak. The hedge along both sides of the hidden roadway crossing the picture is mainly Mexican sunflower with some cypress. The vine at right is ipomoea. A branch of large-leaved cordia is seen at upper left. Grasses here include crab grass, elephant grass, and Guinea grass. The area is cropped to corn, bananas, potatoes, and beans.

Discussion. Human impact is extreme. The vicinity is urbanized with numerous houses, small patches of cultivated land, and many large trees. Telephone lines, such as those carried by the pole at left, and electric power lines cross the area. Camera stations near urban areas were not visited by Shantz and Turner in 1957, and it is likely that they elected not to match the 1920 photograph here. Since the mid-1950's the Meru region has experienced rapid population growth, widespread introduction of cash crops such as coffee and tea, shrinkage of acreage devoted exclusively to livestock, and introduction of new fruits and vegetables providing for a more varied and nutritious diet. One observer of the region has noted, however, that there "are ominous contemporary signs of serious imbalance between population and resources" (Bernard 1993). These signs include deepening gullies, silted streams, forest clearing, and latent malnutrition among children.

BUSHLAND

Although trees are always a conspicuous element of bushlands, shrubs usually dominate the assemblage. The single or layered canopy, with coverage exceeding 20 percent, is no more than 10 m high with scattered taller trees (Pratt and Gwynne 1977). Ground cover is sparse, and wild fires are uncommon. In its extreme form, the bushland thicket, woody plants form a nearly impenetrable barrier to humans or larger ungulates.

A.

B.

C.

Kijabe Town Ia (Plate 23, Study Area 11)

A, (1920). This site, 1 km southwest of Kijabe Town on the boundary between Nakuru and Kiambu districts, is located on the east side of the Rift Valley at the base of the Kikuyu Escarpment. In 1920, Shantz took four photographs here: three panning views from the same position and a fourth from a station a few meters away. This view shows the Rift Valley and Longonot Crater in the distance. Leleshwa is prominent in the foreground as well as Sodom apple and lantana. The dominant grass is Kikuyu grass. The trees at left midground, two of which are dead, are African pencil cedar. Trees and shrubs on the hill at right midground probably include leleshwa, Java lippia, Nyika euphorbia, rhus, and African pencil cedar, all still present in 1992. The lower flanks of Kijabe Hill extend into view on the right and support a grassland with scattered plants of candelabra euphorbia and desert date. As Shantz described: "Lower edge [of forest] marked by secondary bush. Plains extend off as beautiful Themeda grassland with forest extending down in tongues and thickened a little along water and drainage channels" (Shantz n.d., p. 64). The Rift Valley floor approaching Longonot Crater is dominated by shrubland. A single house is visible at center midground.

B, (1957). The foreground, now dominated by the shrub Java lippia, has been disturbed by activities associated with construction of the road, mostly hidden from view by vegetation, that crosses the scene at midground. The conspicuous dark shrub at left is rhus. The dominant grass is Naivasha star grass mixed with common star grass and Kikuyu grass. The hill appears more open than before with more grass and less dense shrubbery. No houses are visible on the valley floor.

C, (1992). The foreground dominants by this date are the perennial grasses common star grass and *Aristida adoensis*, and the annual grasses *Harpachne schimperi,* wiry lovegrass, *Eragrostis braunii*, and crab grass. This composition reflects the advanced stage of disturbance at the site. Isolated plants of red oat grass occur among protective cover of woody plants. On the hill at midground, most of the leleshwa has been cut, each plant producing numerous new stems from its base. The number of trees in the background has increased, although a few, especially candelabra euphorbia and desert date, may persist from 1920. Numerous houses are visible on the fan at the base of Kijabe Hill.

Discussion. The local inhabitants, Kikuyus, cultivate small plots of corn, beans, potatoes and wheat. The bushland at the base of the hillslope shows evidence of cutting, especially of leleshwa, which is used as firewood and in charcoal production (Young and Francombe 1991). No mature plants of this species are evident and most existing individuals have had branches removed in the recent past. Although a vigorous sprouter following fires (Pratt and Knight 1971), the leleshwa shows no evidence of recent burning. Goats and cattle probably roam through all uncultivated areas and indeed their impact is shown by the confinement of highly palatable perennial grasses such as red oat grass and *Hyparrhenia hirta* to protected sites among the shrubs.

A.

B.

C.

Kijabe Town Ib (Plate 24, Study Area 11)

A, (1920). The camera is turned 40° to the right from the previous and does not quite overlap it. The heavily disturbed foreground is dominated by a low shrub cover of leleshwa, *Solanum* cf. *delagoense, Phytolacca* sp. and *Ocimum* sp. (Shantz and Turner 1958, Fig. 59a). Several trees of African pencil cedar, some with pruned branches, dominate at center left. The dense bushland at the foot of the midground slope is dominated by the light-colored leleshwa and other shrubs. A forest of African pencil cedar extends across the base of Kijabe Hill. Although difficult to see in this print, houses may be present just below the hill's summit on the right, suggesting that the hilltop was partially planted with crops by this early date (Shantz and Turner 1958, Fig. 59a). The railroad is 20 years old at this time.

B, (1957). The camera was turned too far left to capture the full view. The foreground has been further disturbed by removal of soil for road ballast. Kikuyu grass, a species favored by disturbance, forms a dense low mat. Weedy shrubs, scattered through this grassy matrix, are leafless at this time but comprise the same species as before (Shantz and Turner 1958, Fig. 59b). The vegetation at midground appears more open with fewer shrubs and more grass than in 1920, and may have been burned since that date. The railroad was shifted upslope in 1949.

C, (1992). The camera position is within a few meters of the earlier locations. The foreground appears to have become increasingly dominated by grasses since 1920 and today supports ruderal species such as common star grass and crab grass. Gone are Kikuyu grass and Naivasha star grass, the dominants in 1957. The large-leaved shrub right of center is Java lippia. The midground slope is once again dominated by a dense bushland comprising leleshwa, *Psiadia arabica,* Java lippia, rhus, *Ocimum kilimandscharicum,* and isolated individuals of both African pencil cedar and pepper tree. The dense patch at the extreme left midground is African pencil cedar. A single candelabra euphorbia stands just left of center.

Discussion. The roadway crossing just below center appears to have largely closed over. The railroad right-of-way has become overgrown since its abandonment after 1949, but is still used as a footpath by the local Kikuyu. The forest of African pencil cedar extending across the base of Kijabe Hill has decreased in size by about 75 percent since 1920. The remaining forest is on steep and broken ground, possibly sparing it from cultivation. The open area below the forest belt has been cultivated in the past but is now returning to bushland dominated by leleshwa, *Psiadia arabica,* and other shrubs. The open vegetation around the summit of Kijabe Hill is heavily cultivated. Scattered houses are visible on the slopes. The forest seen in 1920 along the ridge to the right of the summit is less extensive in 1992.

A.

B.

C.

Kijabe Town Ic (Plate 25, Study Area 11)

A, (1920). The camera was rotated to the right for this view, not quite overlapping with the preceding plate. The dense grass cover in the foreground is Kikuyu grass and common star grass. The shrubs are leleshwa, lantana, and Sodom apple. The railroad dominates the midground. On both sides of the right-of-way, a dense bushland of leleshwa, lantana, and *Psiadia arabica* is found. Beyond the track are seen large trees of African pencil cedar and some cypress plants. A few houses of Kijabe Town are visible on right. A dense forest of African pencil cedar covers the lower slopes of the Kikuyu Escarpment, with higher slopes appearing partially cleared.

B, (1957). As with other views from this same position, the site is grassy but studded with tall forbs that are dry at this season (January). The grasses are Naivasha star grass and common star grass. The wilted forbs are an abutilon, plants with black stems are acalypha, and an occasional Sodom apple can be seen. Most of the large African pencil cedars at midground have been removed. The railroad was shifted farther upslope in 1949. Rectangular fields, clearly outlined by the bordering hedgerows, now replace much of the forest at the base of the escarpment. Farther upslope, the forest is dominated by African pencil cedar.

C, (1992). The camera station is now at the edge of a cultivated field, interplanted to corn and beans. The foreground grasses include common star grass and crab grass. African pencil cedars, although still present, are not as large as some seen in 1920. Tasmanian blue gum (large plants on right) has been planted since 1920. The houses of Kijabe Town, about as numerous as before, are constructed differently. The rectangular cleared areas from the 1957 view are still discernible but mostly support bushland at some stage of recovery from clearing. Farther upslope considerable clearing has occurred in the past, but forest regrowth is evident now. The forested slopes are part of a protected forest plantation where Mexican cypress has been planted with the native African pencil cedar.

Discussion. The trees of African pencil cedar that were conspicuous elements in two of the 1920 views have been removed. The foreground vegetation appears to have changed from a shrubland in 1920, to shrub-perennial grass in 1957, and to annual grass by 1992. Kikuyu grass, noted in 1957, is now largely absent except in the protection of shrubs where it is joined by other perennials such as red oat grass, *Hyparrhenia rufa*, and *H. hirta*. These grasses, nearly extinct here now, have been replaced by the annual *Harpachne schimperi* and the perennial *Aristida adoensis*. These last two species, especially *Aristida*, are common in heavily grazed pastures. The forbs *Ocimum lamiifolium* and Sodom apple are also present and indicate persistent heavy use of the site. Midground vegetation on Kijabe Hill became more open with fewer woody plants during the 1920-1957 period, but the trend was largely reversed during the last 35 years. The top of Kijabe Hill, which is more heavily cultivated than the footslopes, has fields of corn, beans, potatoes, and wheat, and more homesteads than in the past. The increased number of houses and fields in these views is evidence of the population pressures felt throughout the countryside.

Kijabe Town II (Plate 26, Study Area 11)

A, (1920). This camera station is a few meters west of the previous camera station. A dense foreground sward of Naivasha star grass and Kikuyu grass gives way to leleshwa and *Solanum* at midground. The junipers are the same plants as those in Plate 23a. The railroad right-of-way crosses the midground slope. Other midground details are essentially the same as for Plate 23a.

B, (1992). The camera is placed farther forward than before because the original location is now in a field cropped to corn and beans. The dense foreground grass cover is composed of the same species as noted in Plate 23c. The foreground area has been greatly disturbed through earth-moving activity.

Discussion. The foreground has lost most of the shrub cover present in 1920, and the dense grass cover has undergone the same qualitative change noted at the previous station. Midground changes are also consistent with those of the previous station where the 1920 shrubland with scattered African pencil cedars was apparently cleared and has partially recovered by 1992. Most of the observed vegetation changes are the result of human activities. More large trees and houses are apparent on the valley floor in 1992 than in 1920.

WOODLAND

Woodlands support mainly deciduous trees of intermediate stature (up to 20 m in height) with a canopy of more than 20 per cent cover and of branches that may be open or continuous but not thickly interlaced. Shrubs may be absent, or if present contribute no more than 10 percent of the total cover. Grasses and other herbaceous plants dominate the understory, providing the fuel for periodic dry-season fires. Coppice woodland results when the majority of the trees are resprouting after cutting or burning (Pratt and Gwynne 1977). By the time of Shantz' early visit to these woodland sites, many were already under cultivation.

Mashoti Station (Plate 27, Study Area 23)

A, (1920). Located at railroad kilometer marker 49.6, this camera station is about 3 km southeast of Mashoti Station and about 2 km north of Taita Hills Lodge. View is across the plains west of the Taita Hills. Shantz referred to the woodland seen on the low hill in the background as acacia grassland (Shantz n.d.). The railroad track, just out of view on the right, makes a long sweeping turn to the left and crosses the field of view just below the crest of the hill. Its course is marked by the light-colored utility poles in the left background. The dense cover of grass in the foreground is primarily Guinea grass. At left foreground are seen *Maytenus putterlickioides* and a young *Grewia hexaminta.* Many of the trees have been cut and some branches have subsequently produced dense regrowth. A long knife or panga, not well suited to cutting large stems, was used to cut the trees, so many of them are not severed at ground level. *Cordia gharaf* is at far right.

B, (1992). The camera is within a few tens of meters of the original position. The railroad right-of-way has been shifted uphill from its former location. The long sweeping curve of track is still visible across the far hill. The foreground is dominated by Guinea grass, as before, and *Chloris roxburghiana.* The tall shrubs to the right and left of the utility pole are three-thorned acacia. The grassy foreground appears little changed, although the shallow, midground valley has much more woody growth now. The tree species now growing here and on the hill at back include umbrella thorn, Egyptian thorn, three-thorned acacia, and desert date. Most of the trees near the camera station are shrublike clusters of regrowth sprouts. The density of woodland vegetation on the hill appears little changed in 72 years. Other plants seen at the site are *Grewia hexaminta,* Sodom apple, and *Indigofera schimperi.* Red oat grass is the most common grass on the far hill, but additional grasses found with it and not noted above are Naivasha star grass, tanglehead, and African fox tail.

Discussion. Most of the trees, especially umbrella thorn, are being selectively cut for charcoal production. The tree density of this woodland has been little affected by wood harvesting over 72 years. The site is grazed by domestic livestock and by wild herbivores. The abundance of red oat grass, which germinates in response to fire, suggests that the area has been burned within the last few years.

Bura Station (Plate 28, Study Area 22)

A, (1920). The camera, looking toward the southern escarpment of the Taita Hills, is located next to the Kenya Railway station at Bura. The foreground is ". . . more or less ruderal . . . with Solanum, sisal, etc. but not in back. A beautiful mountainous country here. Bananas show on mountain sides . . . . The mountain sides in cultivation--native agriculture not limited as in Europe by lay of the land" (Shantz n.d., p. 161). *Solanum* and agave are in the foreground. A few house roofs are visible at midground.

B, (1992). The camera station is located within a few meters of the original position. The native broad-leaved forbs that dominated the foreground in 1920 have been replaced by Naivasha star grass and Guinea grass. Plantings of corn, beans, cassava and paw paws are evident. At midground, rooftops of shops in Bura are visible along with such exotic trees as casuarina, Red River gum, and neem tree. Also present are the indigenous trees umbrella thorn, Egyptian thorn, and candelabra euphorbia. The steep hillslope and hill summit appear free of cultivation; the slope is dominated by curse of India, candelabra euphorbia, and Egyptian thorn; the summit by silky oak and species of *Croton* and *Eucalyptus*. The forest, still silhouetted against the sky, is somewhat denser than in 1920.

Discussion. Human activity at the site has led to the introduction of fast growing exotic trees such as silky oak and species of eucalyptus. Most of the indigenous trees have been cleared to make way for cultivated crops. Yet the background hillslope still supports a number of indigenous trees, especially *Lannea stuhlmannii* and umbrella thorn.

A. 

B. 

C.

Taita Hills. Shantz visited this small mountain range in 1924 and 1957. Five photographs first taken in 1924 and three in 1957 are presented here.

Taita Hills I (Plate 29, Study Area 20)

A, (1924). This station is 5 km (by air) northwest of Mwatate along the Bura-Wusi road. The view is across the east face of the mountain range. The foreground plants include the indigenous shrubs castor oil plant and triumfetta, as well as non-native trees such as a cypress species and silky oak. The photograph shows scattered homesteads in the valley below. Except for the houses at midground, the dwellings are circular with conical thatched roofs, a common type of construction in the area.

B, (1957). The camera has been shifted slightly to the left. From the new location, at least one of the original midground buildings can still be seen. The photograph is taken from the yard of a mission. A few more buildings occur in the valley below, some with rectangular floor plan. Many fields of corn, pigeon peas, and beans are present as are plantings of bananas. The number of trees is greater than before.

C, (1992). The original camera station is in the yard of the Murray Girls High School. Because a luxuriant growth of foreground plants now blocks the view, the camera position has been relocated just downslope from the buildings seen at midground in the preceding views. Over the foreground patch of corn can be seen guava and, at extreme right, jacaranda. A few strangler fig and bridelia trees are scattered over the slope adjacent to and below the camera. In the valley below and on the hillslope beyond can be seen Cook's araucaria, silky oak, eucalyptus, red-hot poker tree, mango, bridelia, guava, and strangler fig. Shrubs and forbs in the vicinity of the camera station include Mexican sunflower and rhus. Grasses at the site include elephant grass, Guinea grass, and Kikuyu grass. The background ridges and scarp bases have been planted with silky oak and eucalyptus. These and the native croton are the dominants in the scattered stands of trees at the higher elevations.

Discussion. The area is heavily settled and cultivated. Among the homesteads, schools, and shops are fields planted to corn, beans, bananas, avocado, guava, and cassava. Mangoes are common throughout. Although even the steepest slopes are cultivated, terraces planted to elephant grass protect them from excessive erosion. Most of the indigenous trees have been cleared to make room for cultivated gardens, although natives such as strangler fig and bridelia occur in isolated places. The area under cultivation appears to have changed little since 1924. During the same period, however, the number of dwellings has increased roughly threefold, with the greatest increase occurring since 1957.

Taita Hills II (Plate 30, Study Area 20)

A, (1924). Located along a section of abandoned road upslope and around the crest of a ridge from the previous site, the view shows a ridge top in the Taita Hills. The plants in the foreground include red-hot poker tree (extreme right), bark cloth fig, and banana. The facing slope is almost entirely cultivated, with banana the main crop. No houses are present at this early date.

B, (1992). Shrubs in the foreground are Mexican sunflower and curse of India. The slope and the ridge crest are dotted with homesteads and the slope remains under intensive cultivation. The crops are mainly corn, bananas, and beans. Trees visible on the slope include the exotic species silky oak, mango, Tasmanian blue gum, and cypress; and the indigenous species white thorn, melia, red-hot poker tree, and bark cloth fig. Shrubs and forbs seen here include rhus, common rhus, guava, sand olive, lantana, Java lippia, and Mexican sunflower. The most prominent grasses include elephant grass, Kikuyu grass, and tanglehead.

Discussion. Although not widely used nor well executed in the area in 1924, terracing is widely employed today. The 1992 terraces are carefully laid out, closely following contours. Borders are planted to elephant grass. Despite these actions, erosion scars are evident along some of the drainageways at left center. Several trees appear to carry over from the earlier date, most notably the large bark cloth fig at extreme left below center page. That species and red-hot poker tree are common along drainageways while silky oak and species of eucalyptus have been planted among the fields.

Taita Hills III (Plate 31, Study Area 20)

A, (1924). Located a short distance south along the same abandoned road from the previous station, the camera looks across a small valley toward Mwatate, hidden behind the far ridge. The shrubs with large leaves on the steep foreground slope are probably red-hot poker tree. Grass cover is sparse, exposing the foreground rocks. Except for the area near the homestead at lower right, the steep slopes are mostly uncultivated. Small patches may be in various stages of recovery from earlier cropping with faint signs of terracing still apparent. The vegetation has the aspect of wooded grassland.

B, (1992). Woody plants on the foreground slope include red-hot poker tree, guava, rhus, species of cypress, and lantana. Mexican sunflower is in the immediate foreground. Many homesteads are scattered over the far slope. See the listing for the previous site for additional plants.

Discussion. There is evidence of a change since 1924 in the use of the slope across the valley. Dominated by pasture land before, the area is now mainly occupied by small cultivated plots. More houses and trees are also evident. The indigenous white thorn forms dense copses at several locations and is joined by scattered plants of silky oak, Tasmanian blue gum, mango, and a species of cypress.

Taita Hills IV (Plate 32, Study Area 20)

A, (1924). Along the Bura-Wusi road, the view is toward a steep hill and rocky peak beyond. The steep slope is heavily cultivated with bananas, sugar cane, cassava, arrowroot, and Irish potatoes. The trees are broad-leaved croton and bridelia; the shrub is curse of India. One house is seen at upper right and perhaps others at upper left.

B, (1992). The camera position is in the yard of a small kiosk along the road above what is now the Murray Girls High School. The new station was moved several meters to avoid heavy plantings blocking the view from the true position. Rhus is the shrub at right foreground. The large trees include mainly silky oak and eucalyptus, with fewer numbers of jacaranda, red-hot poker tree, guava, mango, bridelia, broad-leaved croton and croton. Several ferns were seen here, including bracken. Grasses at the site include Kikuyu grass, common star grass, and pyramid dropseed. Houses are visible at several levels on the slope.

Discussion. Plantings of banana have changed little while more land has been planted to local food crops such as corn, beans, and potatoes. Trees, which are more abundant now than earlier, are mostly exotic species.

Taita Hills V (Plate 33, Study Area 20)

A, (1924). The foreground shrub with large leaves is red-hot poker tree. The slope is divided into small plots where bananas, corn, and sugar cane are grown. No houses are visible.

B, (1992). The camera station is near the Murray Girls High School entrance. The base of the opposite hill is blocked from view by foreground plants of *Ocimum basilicum* and *Vernonia cinerascens.* The tall, slender tree with light bark at left foreground is jacaranda. Houses, widely scattered among the cultivated fields, can be seen on the slope facing the camera. Trees are more abundant now than in 1924 and include mango (rows of dark trees descending the slope), bark cloth fig, red-hot poker tree, silky oak, Tasmanian blue gum, bitter albizia, and bridelia. A planting of large cypress is seen along the skyline at upper right. Other common plants growing at the site are Mexican sunflower and guava. Grasses here include Guinea grass, Kikuyu grass, and common star grass.

Discussion. The amount of land under cultivation may have changed little over the past 68 years, but judging from the increase in dwellings, the land is more heavily used now. The practice of terracing the fields is universal and goes far in controlling soil loss from the steep slopes. Footpaths connect the houses with the road at upper right, the only thoroughfare in the vicinity for motor vehicles. Human influence on the hill at left appears slight judging from the high density of indigenous white thorn and bridelia trees.

Mwatate-Wundanyi Road. When Shantz and Turner went to the Taita Hills in 1957 to revisit camera stations established by Shantz in 1924, they took some photographs that were not matches of earlier ones, including this series of three overlapping views from a roadside overlook. The camera station for this series is 11.8 km from Mwatate on the Mwatate-Wundanyi Road.

Taita Hills VIa (Plate 34, Study Area 20)

A, (1957). The view is east with the Sagala Hills visible in the left background. The town of Voi is located at the foot of that range just off camera at extreme left. The grass at left foreground is elephant grass. The midground slope is planted to sugar cane and bananas.

B, (1992). The camera was shifted slightly from the original position because heavy vegetation blocked the view. The steep slope immediately below the camera station is settled but with fewer houses than in the valley at midground. The ridge on the left has more trees than earlier. This small forest includes the introduced silky oak, Mexican cypress, Cook's araucaria, and Tasmanian blue gum, as well as the indigenous trees, *Grewia trichocarpa,* African pencil cedar, Senegal palm, red-hot poker tree, and several species of figs.

Discussion. See Discussion for Plate 36.

Taita Hills VIb (Plate 35, Study Area 20)

A, (1957). Less than half the slope appears to be uncultivated bush. The few trees are probably broad-leaved croton and *Grewia trichocarpa.* Beyond the base of the slope the valley floor is extensively planted to corn.

B, (1992). Scattered cultivated plots on the steep slope just below the camera station are planted to corn, beans, mangoes, and bananas. In the midground valley, the crops are corn, sugar cane, bananas, beans, and many mango orchards. Density of houses has increased conspicuously since 1957. Trees have increased in importance here with native acacias and rhus especially numerous. The background is not much settled and is dominated by natural dryland vegetation composed of three-thorned acacia, hook thorn, and species of *Commiphora*.

Discussion. See Discussion for Plate 36.

Taita Hills VIc (Plate 36, Study Area 20)

A, (1957). The view is to the south. In the background at left is the extensive Taita Sisal Estates that surrounds a ridge with Sembi Peak at left and Zongoloni at right.

B, (1992). Plants seen here in addition to those already noted in Plates 34b and 35b are the trees bark cloth fig and Cape fig; the shrub curse of India; the forb Sodom apple; and the grasses Kikuyu grass and Guinea grass.

Discussion. The changes here are similar to those at the other stations in the Taita Hills. The area's favorable climate and proximity to railroad and highways favored early settlement and the process was well advanced by the time of the first photographs in 1957. During subsequent years increased human population resulted in a change from widely scattered houses, with mainly thatched roofs, to more numerous dwellings of more substantial construction. The fields cover more total area now than before, and individual plots are smaller in size. The increase in number of trees is typical of mountainous habitats observed elsewhere. An overall decrease in indigenous trees such as *Grewia trichocarpa,* broad-leaved croton, and African pencil cedar is more than offset by an increase in plantings of exotics such as silky oak, Tasmanian blue gum, and Cook's araucaria.

A.

B.

C.

Runyenje Ia (Plate 37, Study Area 7)

A, (1920). The station lies about 2 km southwest of Runyenje along the Embu-Meru highway and overlooks waterfalls on two branches of the Ena River. The large roadside plant at right foreground is a young broad-leaved croton. The grass growing beneath it is Kodo millet. The plants above the road are triumfetta and broad-leaved croton. The ridge at distant midground has been partially planted to fields of corn. Thatched roofs of several houses are visible right of center. The remainder of the hillslope supports scattered trees, probably croton and *Grewia trichocarpa*. This area, in various stages of post-cultivation recovery, is probably dominated by large-leaved cordia, Java lippia (low shrub), and croton. The background hills are not clearly shown but a few trees are visible along the ridge tops.

B, (1957). Two photographs from this year have been spliced together to match the 1920 view. The camera is within a few meters of the original position. The area above the road at left is now covered by a dense sward of grass, probably common star grass and Kikuyu grass, and the shrub curse of India. The trees at the top of the bank are large-leaved cordia and bridelia. The partially "healed" vertical bank alongside the road now supports common star grass and *Aristida adoensis*. The dense growth seen in the foreground in 1920 has been replaced with golden wattle and possibly coffee. The midground ridge has more, and in some instances larger, trees than before, suggesting that the local inhabitants practice some form of agroforestry. Fields of corn are still apparent with somewhat more acreage than before. The steep hillslope appears less open with more and larger shrubs than in 1920. The houses are not visible and may be gone. The background hills appear to support a dense shrub cover with scattered trees.

C, (1992). The road pictured here is no longer maintained. A new paved highway linking Meru and Embu lies a few meters to the left. The hedge at right foreground is a type of cypress mixed with curse of India. Branches of golden wattle are seen at upper right foreground. The large roadside tree just left of center is silky oak. Other trees in view or nearby include a species of acacia, along with croton, banana, and avocado. The bank at left, appearing free of recent disturbance, is less steep and is now covered with elephant grass. Pyramid dropseed is the roadside grass along the base of the hedge at right. The midground slope, no longer visible from this station, is covered with planted silky oak, species of acacia, and flame tree as well as isolated, native broad-leaved croton and bridelia.

Discussion. The vegetation of the area is wooded bushland with many non-native constituents. Shantz noted in 1920 that "sweet potatoes and a tall brand of pea, so common in Congo" were grown at this site. The former subsistence farming has been replaced largely by commercial operations involving coffee, bananas, beans, and corn. The cypress hedge is probably planted but the accompanying curse of India, widely naturalized in Kenya, is increasing rapidly by seed in nearby fallow fields. Although no longer visible from this station, the slopes seen earlier at midground and beyond are more heavily cultivated now than in 1957.

A.

B.

C.

Runyenje Ib (Plate 38, Study Area 7)

A, (1920). This view, overlooking Ndungu Falls on a tributary of the Ena River, is at a point farther right than the one published as Fig. 62a in Shantz and Turner (1958). It is used here instead because of the greater area covered. Bananas, passion fruit, and triumfetta appear in the foreground together with young trees of broad-leaved croton. The white-barked trees at left-midground are white thorn and at center-midground, broad-leaved croton. The dark-crowned trees along the midground gullies are peacock flower. Scattered individuals of bridelia and large-leaved cordia are also visible along the valley floor and at midground. On the hill at left is the Kirimiri Forest, comprising mainly croton and broad-leaved croton with some native olives. The open patches on the hills appear to be in various stages of post-cultivation recovery.

B, (1957). The location is almost exact but the camera is aimed too far left. Behind the cattle, the foreground plants include the trees peacock flower and *Grewia trichocarpa*; the shrub triumfetta; the large forb Sodom apple; and Kikuyu grass. The hillslopes are much more heavily covered by trees and shrubs than before, apparently the result of a long period of no cultivation, the beginning of which may antedate 1920. Some cropping is practiced on the slope at right but less than in 1920. The Kirimiri Forest has advanced down slope on the hill at left into an area of ground already abandoned in 1920. Cattle are not confined to paddocks.

C, (1992). The camera is shifted several meters to the right of the original position to obtain a more or less unobstructed view of the scene. A small village has been built at the camera station since 1957 and buildings and plantings now block the view. Because of the new position, perspective for some features has changed slightly. Crops planted here are coffee, bananas, corn, cassava, beans, and avocados. Hedges are commonly of cypress. The trees on the hillslopes below the camera are mainly casuarina, golden wattle, silky oak, species of acacia, and avocado. Shrubs and forbs found in the vicinity are triumfetta, marigold, and bur-marigold. Grasses include Kikuyu grass, elephant grass, and Kodo millet. The trees in the Kirimiri Forest are mainly silky oak, Red River gum, and croton, a shift in composition since 1920. The crops here are the same as reported for the previous view.

Discussion. This area is now densely inhabited, with numerous homesteads scattered across the landscape. Terraces of the fields, which are more numerous now than before, are protected by plantings of elephant grass along the terrace crests. This grass serves as fodder and as a soil stabilizer. Grazing practices here have changed markedly since 1957, as disclosed by the local inhabitants. Plots of land have been surveyed, marked, and owners designated. Consequently, what was formerly common ground, grazed freely by roaming livestock, is now privately held, and owners confine their livestock to small paddocks and bomas, resulting in zero grazing (Bernard 1993).

A.

B.

C.

Runyenje Ic (Plate 39, Study Area 7)

A, (1920). Shantz took this photograph of the Ena River Falls by turning his camera approximately 40° to the right from the previous view. Bananas, passion fruit, and triumfetta appear in the foreground. The plantings on the steep slopes across the valley include elephant grass, sweet potatoes, and sorghum (Shantz and Marbut 1923, Fig. 8) as well as corn and beans. Two fallow plots support what appears to be a dense cover of grasses (probably crab grass and Kodo millet). The abandoned fields on the steep slope at left are occupied by low shrubs that probably include *Psiadia arabica* and curse of India. Croton, the conspicuous white-stemmed trees near the waterfall, is growing in a small remnant of temperate rainforest (Shantz and Turner 1958, Fig. 61a). On the flat ground at the top of the steep slopes can be seen more cropped areas and, at the right, several houses.

B, (1957). The camera station is slightly too far left for an exact match. The foreground trees are bridelia, bitter albizia, and *Markhamia platycalyx.* The cultivated fields left of the waterfalls have been abandoned for varying periods of time, perhaps at least since Shantz' earlier visit. The fields at far left, both at the top of the hill and on the steep slope below, are still under cultivation. Those on the steep slope are more prominently terraced than before. A road has been built along the hilltop at left. The forest about the waterfall has grown denser, and several of the white-stemmed croton have persisted. The rolling terrain above the steep slopes is more heavily cultivated than before. Many houses are visible at upper right. Because of seasonal low flow, the waterfall, double in 1920, occupies only one of its two chutes.

C, (1992). Fields and houses now occupy the original camera station and the camera was moved several meters to secure a view of the waterfall and terrain across the valley. The plot in the foreground is cropped to coffee with silky oak planted as cover. The fields at far left are planted with bananas, corn, and passion fruit. The tops of the prominent terraces are planted with elephant grass, which controls erosion and is used as fodder for the cattle that are now confined to small paddocks for grazing during the day and to homestead bomas at night. The fields just left of center have been cleared and planted to corn and beans since 1957. There are many new houses on the level ground across the valley, and in the vicinity of the camera station.

Discussion. Although the 1957 view shows many retired fields, this site has seen an overall increase in settlement and cultivation since 1920. The small remnant of tropical rainforest is now almost gone. Most of the indigenous broad-leaved croton has been harvested and replaced with cropland or with fast-growing, introduced trees such as silky oak and several species of eucalyptus. Coffee plantings are more important now, as are fields of bananas, beans and corn. Kikuyu grass, the most abundant grass during earlier times, is now largely confined to the yards of the homesteads.

FOREST

Forests include lands dominated by trees with stems of columnar form, often reaching heights of 50 m or more. These trees, which form a continuous canopy of intertwining and touching branches, are mainly evergreen, except following disturbance when deciduous trees may be prominent. Woody vines are common. Growth of low shrubs and herbs, limited by the dense shade, incompletely covers the forest floor. The few grasses present are limited to broad-leaved species.

A.

B.

C.

Kikuyu Escarpment (Plate 40, Study Area 12)

A, (1920). This camera station is 2.4 km along the road from Kijabe to Kijabe Town. The setting is the steep west face of the Kikuyu Escarpment overlooking the Great Central Rift Valley. The foreground vegetation is heavily disturbed and is dominated by a low, rank growth of *Cussonia* sp. and vines. Slopes above and below the road cut at right appear to have been recently cutover and are now dominated by low shrubs and vines. The taller trees with light trunks are wild olives. African pencil cedar is scattered about the slopes and is probably the dominant tree along the crest of the ridge. Almost 50 percent of the slope appears devoid of trees and is covered by a dense growth following removal of original forest (Shantz and Turner 1958, Fig. 58a).

B, (1957). Although vegetation was described as "much as before" (Shantz and Turner 1958, Fig. 58b) the foreground appears to have taller plants. The area above the road appears to support larger plants and less dense ground cover of vines. The larger trees have been removed and the former patches of shrubs are occupied by taller plants.

C, (1992). This view is atypical of most in this study in that cultural influences on the landscape have diminished rather than increased with time. The plant just right of center is euclea, a common dominant on footslopes of the highlands following bush clearing. The large-crowned trees at midground and to the left are cape chestnut. Cape chestnut, African pencil cedar and wild olive are dominants accompanied by lesser quantities of East African olive. The dominant shrubs are *Dombeya faucicola*, euclea, *Acokanthera longiflora,* and *Hibiscus fuscus*. Grasses, which occupy less than 10 percent of the ground cover, include Guinea grass and common star grass.

Discussion. As Shantz walked from the escarpment summit down to Kijabe he noted, "Most of country second growth--or weed stage--following a slow and deliberate native deforestation. Artificial forests of Eucalyptus, [golden] wattle, and cypress are extensive on the whole escarpment. But original forest is represented here and there by juniper which is very scraggly as a whole" (Shantz n.d., p. 51). Since 1920 there has been a definite increase in large woody canopy species with indigenous *Olea* and *Calodendrum* replacing African pencil cedars along the hilltops. Human impact has apparently diminished, although selective cutting of the forest trees probably continues. Such timber harvesting is controlled by the Forestry Department of the Ministry of Environment and Natural Resources.

RIVERINE FOREST

Several of our views show vegetation typical of riverside habitats. Hardly affected by regional rainfall conditions, these moist habitats often support plant growth that is denser and taller than would be expected for their region of occurrence.

Athi River (Plate 41, Study Area 14)

A, (1920). This camera station is along the approach to the Athi River bridge, 23 km southeast of Nairobi. Taken from a railroad car as it travelled near the Athi River, this photo shows a view across the river (obscured by the heavy growth of trees) and over the plains beyond to the Mua Hills in the background. The riparian tree is fever tree. The dense spherical trees on the plain at right are desert date, probably shaped by giraffe browsing. The herded cattle may be watering at the river.

B, (1992). Because the picture is taken from a tripod on the track bed instead of from a moving railroad car, the camera placement is 2 m to 3 m lower than before. The foreground fever trees have been replaced by a field of corn. Bananas now border the river, which may have a somewhat altered course now compared to 1920. The coarse grass along the gravel roadbed in the foreground is African fox tail. The few trees on the bank across the river are fever trees. The dense row of fever trees beyond the open corn field at midground lines a distant meander of the river. The broad, low shrub at left in the distant field is hook thorn. Pepper tree is growing nearby; *Eucalyptus* sp. at midground. *Capparis tomentosa,* not identifiable in the photograph, grows along the river's edge. Other forbs and shrubs of the immediate vicinity include the indigenous castor oil plant and Sodom apple, and the exotic Indian fig. Other grasses include elephant grass, Maasai grass, Guinea grass, and common star grass. On the background hill occur such native shrubs and trees as white thorn and *Grewia* sp.

Discussion. The immediate vicinity of the camera station is under cultivation, and several buildings (out of view) occur nearby. The riparian strip that skirts the river has been partially cleared for a banana plantation. In addition, corn, beans, cowpeas, and other vegetables are grown nearby. Although seemingly remote from Nairobi, this site lies in the path of city expansion as shown by the numerous heavy-industry complexes already established between here and town.

A.

B.

C.

Thika I (Plate 42, Study Area 9)

A, (1920). This camera station, overlooking falls on the Chania River, is located on an observation terrace at the Blue Post Hotel, Thika, in a region made famous by the writings of Elspeth Huxley. The plant at left foreground is orange-leaved croton. The white stemmed trees at right- and mid-foreground are a species of fig *(Ficus vallis-choudae)*. Other trees that can be identified are peacock flower and markhamia. Shantz draws attention to a planting of eucalyptus on the hill at left (Shantz n.d., field notes, p. 174).

B, (1957). Foreground plants include the introduced Indian fig (left of center) and cotyledon (occupying most of the foreground), and a native aloe. The rock at right foreground in the 1920 view is probably the one just to the left of the aloe at right of center in this view. Vegetation left of the bridge is more open, possibly to make way for the inlet to the water treatment plant that lies just downstream from this site. This plant provides water for Thika as well as Nairobi. The planting of eucalyptus seen in 1920 on the hilltop at upper left has been removed and replaced by coffee. The hillslope appears little changed.

C, (1992). Observers are now restricted to somewhat higher ground than before by a safety railing so that the old camera station was not reoccupied. The prominent foreground rock, still present but hidden by plants, was used to position the camera. The Indian fig from 1957 has been replaced by other introductions also from the Americas: *Agave* sp. at left and curse of India at center (obscuring base of falls). The broad-leaved plant just right of the waterfalls is dombeya and to the right of that is fig. Trees flanking the waterfalls include peacock flower, markhamia, and white-galled acacia. The hillslope is now hidden by the dense tree growth next to the bridge.

Discussion. Except for the clearing of trees on the hilltop and beside the bridge in 1957, vegetation density has remained relatively stable since 1920. The riparian vegetation is on the grounds of the Blue Post Hotel and is maintained in a reasonably natural state.

A.

B.

C.

Thika II (Plate 43, Study Area 9)

A, (1920). The falls are located on the Gitathuru River about 600 m upstream from its confluence with the Chania River at which point the Thika River begins. This site, as the previous, is on the grounds of the Blue Post Hotel, Thika. Identified only as "falls and riparian forest at Thika" (Shantz n.d., field notes, p. 190), this view depicts the forest cover growing along a perennial stream in a grassland region.

B, (1957). Taken farther forward than the original, this view shows that the rocks in the streambed have been relatively stable during the previous 37 years. The two strongly leaning trees seen on the right in 1920 are gone. The water level is lower (January) than in 1920 (May).

C, (1992). The rocks remain fixed in place during the 35 years since the previous photograph was taken, although a large log lies across the prominent rock near the center of the picture. The tree at extreme left, an old specimen of newtonia, is now dead. Other trees in the area are broad-leaved croton, wild olive, markhamia, bark cloth fig, *Ficus vallis-choudae,* and strangler fig. Disturbance on the banks of the near side of the river has promoted a dense growth of curse of India and Mexican sunflower. Other understory shrubs of this dense riverine forest include rhus and acalypha. The two figs, bark cloth fig and strangler fig, are parasitic on peacock flower, killing some. Seedlings of most of the tree species can be found on the forest floor.

Discussion. See Discussion for Plate 44.

A.

B.

C.

Thika III (Plate 44, Study Area 9)

A, (1920). This is a downstream view from a station only a few meters from the previous. The tree at right foreground is adina. The straight-boled trees on the opposite bank are newtonia. Manilakara is scattered along both banks (Shantz and Turner 1958).

B, (1957). Little change is evident since the previous photograph was taken, although the stem diameters of the large trees has increased about 30 percent since 1920 (Shantz and Turner 1958).

C, (1992). Several changes are notable. While adina remains, two of the leaning trees from earlier dates (at center and left of center) are gone. Fewer newtonia are present, but this species is still probably the dominant tree here. The large rock in the stream at center is new and probably rolled down the slope from above left. The rock is covered with a veneer of mosses and lichens similar to that of nearby stable boulders, indicating that it has been in place for a long time.

Discussion. The stability of the riparian vegetation at Thika from 1920 to 1957, noted by Shantz and Turner (1958), did not continue for the next 35 years. Stem counts on the photographs taken at streamside show a decline in numbers of trees along the river banks by 1992. Several trees have died and are still standing, suggesting that the decline is at least partly due to natural attrition.

Tana River at Sagana I (Plate 45, Study Area 8)

A, (1920). Taken from a bridge over the Tana River, Shantz moved less than a kilometer from the site shown in Plate 20 to take this upstream view. The large tree at right is bark cloth fig. The adjacent uplands are uncultivated shrubland.

B, (1992). Bark cloth fig is still visible at right. The large tree at center is Cape fig. Other plants of the riparian strips bordering the river are adrue, curse of India, and *Phragmites mauritianus.* Red-hot poker tree, bridelia, and Nyika euphorbia are large indigenous plants visible from this point. The introduced species silky oak and finger euphorbia are also visible here.

Discussion. Curse of India now dominates the banks of the river. As is typical of the entire district, increased cultivation of the uplands has occurred since 1920, and they are now extensively planted with fields of corn, bananas, mangoes, and beans.

Tana River at Sagana II (Plate 46, Study Area 8)

A, (1920). This camera station is just a bridge width from the preceding, with the view looking downstream. Nyika euphorbia, at left, is the most prominent plant in this view. Although certain identification of many plants is not possible, some, such as the following, are distinctive enough to name: Cape fig (tree with pale bark on river bank at left), bark cloth fig (dark tree on bank at midground), *Achyranthes aspera* (the dominant herbaceous understory), and *Cissus rotundifolia* (the vine). The uplands at midground appear not to have been cultivated in the recent past.

B, (1992). The camera is shifted to the right because the exact position is now blocked by a large strangler fig. Behind it is Cape fig and silky oak. The large round tree just right of center is *Teclea* sp. Mexican sunflowers are established on the river bank flanking the *Teclea* and are joined by *Phragmites mauritianus,* adrue, and antelope grass. The valley floor beyond the river is planted to corn, bananas, mangoes, and beans.

Discussion. Conditions are much the same as in the previous set of plates although here the dominant shrub along the river bank is Mexican sunflower rather than curse of India.

A.

B.

C.

Thuchi River (Plate 47, Study Area 6)

A, (1920). Several photographs were taken in 1920 from or near this position which is on a bridge over the Thuchi River. The only one matched in 1957 is used in our report. Shantz describes the forest here as "a dense growth of lianas and trees such as *Piptadenia* sp., *Xymalos monospora, Adina* sp., etc. Wild bananas are *Musa* cf. *fecunda.* The herbaceous cover along the stream comprises a dense growth of *Melanthera scandens, Impatiens* sp., *Basella rubra,* and *Bidens* sp." (Shantz and Turner 1958, Fig. 63a).

B, (1957). The foreground rock in the stream is the same as in the previous. The streamside vegetation and forest beyond appear little changed since 1920, although "undergrowth and lianas apparently less than is shown in the earlier picture" (Shantz and Turner 1958, Fig. 63b). Several of the white-stemmed trees have persisted.

C, (1992). The scene is much changed, with more open growth than before. The former streamside shrubs have been replaced by *Phragmites mauritianus* and *Plectranthus barbatus.* Just back from the stream's edge can be found plantings of corn and bananas. The forest is gone, replaced by curse of India, silky oak, and eucalyptus. The steep slopes are mainly planted to tea and coffee with lesser amounts of banana, corn, and avocado. Out of view, on the more level terrain beyond the slopes of the valley, are numerous habitations scattered among fields of beans and corn.

Discussion. Although the bridge from which the photographs were taken still exists, the road it serves has been abandoned since 1983. According to one local inhabitant, the people of the region did not use this small valley in earlier times through fear of wild animals, especially leopards and cheetahs. As cultivation and homesteads encroached on the forest, these animals left, although leopards may still be seen. The valley and uplands are now heavily settled and even the steep slopes are under intensive use. The steep slopes are terraced with ridges planted to elephant grass as a soil binder. The small patch of native forest viewed by the camera has been almost completely replaced by the exotics noted above, although nearby can be found trees such as *Albizia* sp., broad-leaved croton, large-leaved cordia, and bridelia. Large trees of *Albizia* sp. growing along the approach to the river were being felled and cut into lumber at the time of the last visit. The inhabitants regard these trees as an important source of timber when fully grown. When felled, little is wasted--the branches being converted to charcoal and the trunks made into lumber.

CULTIVATED OR DISTURBED LAND

Two early views record conditions in habitats that wer e already heavily impacted by human activity with evidence of original vegetation destroyed.

Mwatate (Plate 48, Study Area 21)

A, (1920). The camera station is in the town of Mwatate, approximately 0.8 km southwest of the intersection between the Voi-Taveta road and the road to Wundanyi. The view shows Longoloni Hill at left. The picture is blurred, presumably because it was taken from a moving train. Shantz described this view as "Out across river bottom with bananas, sugar cane and over plantings of trees and grass" (Shantz n.d., p. 162). The heavily cultivated area of the foreground and midground gives way to more fields with intervening hedgerows on the low ridge beyond.

B, (1992). The view from the original position is blocked by plantings so the camera was relocated upslope and to the left. The new position is between the railroad track and the dispensary at Mwatate and lies among cultivated fields of corn, beans, pigeon peas, paw paws, and mangoes. The sloping ridge at midground is planted to scattered trees of tamarind and mango which are interspersed with the indigenous *Lannea stuhlmannii, Sterculia rhynchocarpa,* umbrella thorn, fever tree, and baobab. Not mentioned by Shantz and perhaps not then in existence are the fields of sisal that now extend all across the view in the background. The light patch left of center is a large area with fiber-drying racks. To the left of that is a field being irrigated with a sprinkler system. Grasses here include Kenya needle grass, Guinea grass, and tanglehead.

Discussion. Because of the shift in camera position, about the only valid comparison that can be made between the two views concerns the low midground ridge visible in both. There has been remarkably little change in 72 years. The area is cultivated to roughly the same extent as before. The trees, although larger now, appear about as numerous as in 1920.

A.

B.

C.

Voi (Plate 49, Study Area 19)

A, (1920). View is toward the north end of the Sagala Hills as seen from the railway station at Voi. Shantz probably alighted from the train to take this photograph. The foreground is disturbed and supports a growth of such ruderal plants as *Justicia flava, Achyranthes aspera, Trichodesma zeylanicum, Hypoestes triflora,* and Guinea grass. The tree at upper left is Cape fig. At midground the larger dark trees are Cape fig and the smaller are *Acacia nubica* and umbrella thorn. The slopes of the Sagala Hills are dominated by a rather open dryland vegetation, poorly shown in this flat print, with the trees along the skyline probably being hook thorn, Egyptian thorn, and *Boscia angustifolia.*

B, (1957). The camera has been moved to the right a few meters where the scene is dominated by houses occupied by families of railroad employees. The foreground grass is Naivasha star grass and some planted corn. The fig trees present earlier have been removed to make way for the houses, and trees of *Cassia siamea* and neem tree are now present just beyond. The fan at the base of the mountain may have been partially cleared. Changes on the hillslopes are difficult to discern but some of the trees along the skyline have been removed.

C, (1992). The camera is positioned to duplicate the 1957 location, thus avoiding a building and a large Cape fig that otherwise block the view. The dense grassy sward in the foreground is Naivasha star grass. The trees nearest the camera station are planted and include the exotics *Cassia siamea* and neem tree, and the indigenous Egyptian thorn. At midground a plantation of mango is evident as are naturally occurring Cape fig and baobab. Other trees present between the camera and the steep slope are tamarind, three-thorned acacia, hook thorn, and Tasmanian blue gum. The shrub *Acalypha fruticosa,* and African fox tail grass, are also present nearby. The dryland vegetation on the steep slope at back includes hook thorn, three-thorned acacia, three species of *Commiphora* growing with an understory comprising in part Kenya needle grass, *Enteropogon macrostachyus,* African fox tail, and hairy herringbone grass. Trees visible along the ridgeline include several acacia species, croton (stunted), Natal rhus, and common rhus.

Discussion. The photograph station is now heavily settled by the railroad workers' families. The fan and lower mountain slopes are dotted by homesteads and are probably subjected to livestock grazing, firewood collection, charcoal burning, and selective cutting of construction timber. Less valuable trees such as *Commiphora* are abundant. The upper steep slopes of the mountain probably are little affected by the proximity of a town.

Patterns of Change

The patterns of change noted at the camera stations represented by 49 sets of replicated photographs spanning 35 to 72 years can be characterized as follows: (1) on rangelands in the most arid areas visited, woody species have increased, grasses have usually declined and, at some arid stations, the dominant shrub species of the 1920s has been wholly replaced by another shrub today; (2) in semiarid areas only marginally suited for cultivation, land has been newly cultivated and associated dwellings built, although some new fields have been abandoned recently; (3) in the more humid highland areas, acreage under cultivation has remained the same or increased, usually with a decrease in plot size but with an accompanying increase in the number of houses and cultivated trees; (4) forested lands support more trees now than earlier and the new trees are usually fast-growing introduced species. These changes are described in more detail below.

Changes in semiarid areas

None of the sites visited represent truely arid conditions. The slightly more moist conditions included here under semiarid include precipitation values from 340 mm to 1050 mm. The sites in the semiarid category are cultivated or used as open rangeland for grazing by domestic or native herbivores.

Cultivated lands. Plates 5, 13, and 14 are examples of semiarid grassland sites that have been successfully converted to cropland. At others, attempts at cropping have been less successful (Plates 16-19). Because of the relatively low rainfall in these areas, crop production is feasible but risky. The gamble farming that is practiced fails in times of drought, resulting in abandonment of cultivated areas and degradation of uncultivated adjacent tracts through wood cutting, heavy grazing, and trampling. These practices lead to loss of topsoil during periods of intense rainfall, resulting in even greater risk of crop failure during bad times (Sindiga 1984). In 1920 these lands were largely unoccupied; 72 years later many such areas were settled as shown by photographs taken between Isiolo and Meru (Plates16-19).

Rangelands. Changes at 13 rangeland sites involved increases in woody plants, usually with accompanying decline in grass cover (Plates 1-13).

The acacias most often found in increased numbers were umbrella thorn, whistling thorn, false umbrella thorn, and three-thorned acacia. Whistling thorn, a plant sensitive to fires and highly palatable to giraffes, has been used as an indicator of the impact of these two forces on lands where the plant occurs. In Nairobi National Park whistling thorn tends to remain small (because of the heavy browsing by giraffes) and to become denser (because of the absence of fires) (Dagg and Foster 1976). These dual forces may have shaped the populations of this plant seen in Plates 6, 11, and 12. Many of the acacia plants in these views show the "inverse browse line" that is induced by giraffes when they browse from above in contrast to the high browse line found on tall trees that have been browsed from below by these long-necked animals (Dagg and Foster 1976). Plate 13, by contrast, appears to show an acacia population that is free of giraffes (indicated by the large stature of the plants) and fires (indicated by the large number of plants). When in effect, the dual impact of these forces may be strong enough to convert bush into grassland (Dagg and Foster 1976).

Changes in the humid highlands

Landscapes in the humid highlands that appear in our photographs fall into two classes: cultivated land and forests.

Cultivated lands. Most of the camera stations located in the moister areas of Kenya show increased area under cultivation and a concomitant increase in tree density. Good examples of these small landholder tree plantings are seen in Plates 25, 28-39, and 48 where they occur as woodlots, hedgerows, ornamental and shade plantings, serving as a source of timber, fuelwood, and food. Some species, such as golden wattle, have multiple use, supplying poles for construction, wood for fuel, and bark as a cash crop. As pointed out for the Machakos District (Tiffen et al. 1994), the increase in tree density is remarkable, occurring despite a fivefold increase in the human population. At this stage, there is no evidence that tree harvesting from these small plantings exceeds tree replacement rate.

Although no change is seen in cultivated land in one (Plate 29), 21 photographs show an increase through time in cultivated acreage (Plates 21-26, 30-36, 38, 39, 41, 45, 46). Terracing, an uncommon feature in the early views, is a common feature in the most recent ones. The increase in cultivated land seen in these photographs is consistent with findings for the Machakos District where increases in land under cultivation followed closely the increase in human population of the area. Terracing also became a universal practice on the area's cultivated slopes (Tiffen et al. 1994). Many of the 1992 photographs of cultivated areas give the impression of heavily utilized but well-managed landscapes. (See especially Plate 38c).

Forests. By the time of Shantz' first visit, Kenya's limited forested areas were under considerable pressure from humans. The native forests had been harvested without the goal of sustainability in mind. Locally important as sources of construction material and fuel wood, the forested areas were targeted during this time for urgent government-sponsored afforestation projects to reverse the downward trend. Another reason cited for afforestation was the perception, common in the 1930s, that these upland plant communities had a positive effect on rainfall (Tiffen et al. 1994). The process of establishing (gazetting) reserves has been described in detail for the Machakos District (Tiffen et al. 1994) and probably applies in large measure to other areas. Potential government-run reserves, found mainly on hilltops, were first designated for reserve status by local native councils. Then any people cultivating or grazing in the area were evicted. An intensive afforestation program, mostly involving exotic trees, was next undertaken, so that ultimately the "native forest reserves" were actually government-run plantations of non-native species of eucalyptus, pine, silky oak, and cypress. These new forests are carefully tended today at government expense and, when mature, the trees are harvested under government supervision.

Some variant of the Machakos District scheme for establishing government reserves was used throughout the highland areas and was probably in its early stages at the time of Shantz' first visit. Examples of this form of afforestation are seen in the Kirimiri Forest near Runyenje (Plate 38) and on the Kikuyu Escarpment near Kijabe Town (Plate 25). An instance of natural afforestation where native species have recolonized previously cleared areas is also seen on the Kikuyu Escarpment near Kijabe (Plate 40). Natural afforestation is probably representative of but a small fraction of the forests in the vicinity, judging from Shantz' observation in 1920 that most of the native forest here had been removed and replaced by "artificial" forests.

Causes of Change

Vegetation change in Kenya has been greatly influenced by the growth of the country's human population. Set at 3.5 million in 1900 (McEvedy and Jones 1979), Kenya's population today, with a doubling time of 19 years, is one of the fastest growing in the world. Its 1992 population of 26.2 million is projected to reach 62.3 million by 2025 (Population Reference Bureau 1992).

Although the problems associated with increasing numbers of people are most pronounced in urban areas, the impact of rapid population growth is also marked in the countryside. As the population expands, lands of high production potential, such as those within high rainfall areas also served by modern transportation systems, are fully exploited first. The greatest population changes have been observed in the central highlands where population expansion has prompted a major rural-to-rural migration in recent decades (Odingo 1971). As the pressures build in the more favored regions and the migrants move down the ecological and elevational gradient, they find that the old familiar agricultural systems are poorly adapted to the new situation. Food production suffers as a result. Old patterns of shifting cultivation, for example, involving long periods of rest following cropping, have broken down under the pressure of increased population and of land adjudication (Anonymous 1990, Bernard 1993). At the low end of the gradient, extensive areas of bushland have been converted to farmland.

By the 1920s our photographs show large areas already under cultivation at many highland stations (e.g., Taita Hills [Plates 29-36] and Runyenje [Plates 37-39]). As the population multiplied, more houses were built to accommodate the growing population, and fragmentation of the highland farms increased. Perhaps some people were forced into semiarid areas such as those represented at camera stations between Meru and Isiolo (Plates16-19). (In the third of a century between 1948 and 1982, for example, cultivated land in the lower Embu/Meru district increased from 12 percent of the area to 41 percent [Anonymous 1990]). The foregoing landscape modifications seem to be the result of direct or indirect increases in human populations. What influence, if any, has climate had on the observed changes?

For the highlands stations with long records, the period from 1920 to 1945 seems to have been a time of extreme droughts that occurred with greater frequency than just before or just after that period (Fort Hall, Embu, Meru, Machakos, and Makindu). Voi, on the lowlands nearer the coast, shows a similar period but from about 1910 to 1935.

The most dramatic changes not attributable to artificial clearing or planting are seen at the stations north of Isiolo where grasslands of bamboo grass and Maasai grass are now forests of acacia (Plates 18 and 19). What are the possible causes of such changes? If climate change is involved in the shift from grassland to acacia forest, this would presumably require a substantial increase in rainfall. The climate records do not show such an increase. Indeed, the opposite appears to be the case. The records for Isiolo, which begin in 1930, show a long period of two decades from 1931 when rainfall was mostly less than the long-term average (Appendix 6). The records for Meru show that severe drought occurred during the season of long rains about 25 percent of the time from 1920 to 1945, followed by a period when severe drought during that season occurred but twice from 1946 to 1973 (Appendix 6). The quarter century of repeated severe droughts prior to 1945 may have contributed to the vegetation changes by producing a weakened grass cover with a resulting reduction in fires. Woody plants that may have become established during intermittent wet years would survive the dry ones. With diminished competition, the newly established seedlings would prosper during subsequent wet years. A similar scenario has been proposed to explain the expansion of semiarid juniper woodlands in North America (Miller and Wigand 1994), except that grazing by domestic livestock is an essential component of the changing system there. The likelihood is great that the same condition applies in the Isiolo-Archer's Post area and that without pressure from domestic livestock, the complex interactions between fires and native browsers and grazers would have maintained a grassland free of a woody component. In addition, the decline in large browsers such as elephants, a group formerly represented in North America by mastodonts and other proboscideans, has probably accelerated the trend toward increased woody plants. The full impact of native herbivore losses in areas surrounding national parks in Africa is of growing concern. For example, during the five-year period, 1988 to 1992, resident ungulate populations declined by 95 percent outside the western Serengeti boundary (Campbell and Borner 1995). These losses among native herbivores are often accompanied by increases in livestock and human populations.

At Ol Doinyo Sabachi (Plate 1), where more palatable shrubs have been replaced by less palatable ones, the change is also most likely related to grazing history. In that instance, and those near Archer's Post, all of which show increased woody plant growth, a climatic trend does not seem to explain the change.

Most early studies of rangeland dynamics took place in North America, where woody plant encroachment into grasslands has been a classic symptom of grassland decline. Vegetation changes on the Kenyan rangelands might not be expected to follow the same pathways as those on their American counterparts for several reasons. First, the short North American roster of large herbivores, led today by domestic cattle, is far exceeded in Kenya by the kinds and numbers of wild herbivores and domesticates utilizing the rangeland. Second, a comparable path is unlikely because of the diverse populations of pastoralists that still inhabit much of the nonurban, uncultivated land of Kenya. Their unique dietary preferences for milk, blood, and meat are reflected in the way each group manages its herds of domestic livestock. These systems of animal husbandry are often followed with little effort to maintain livestock numbers in equilibrium with range potential. A third important agent contributing to a possible distinctive rangeland dynamic in Kenya is the occurrence of animal diseases. Although afflictions such as Texas fever were a potent force during the early period of ranching in North America (Osgood 1929), advances in veterinary medicine have reduced their influence to a negligible possibility today. Similar reductions of disease have been noted for the Serengeti (Pellew 1983), where effective veterinary practices during the last four decades have greatly diminished the effects of many livestock diseases. High cattle numbers have resulted, which in turn have promoted low fire frequency and bush encroachment. However, outbreaks of African diseases, such as anthrax, rinderpest and trypanosomiasis, still remain distinctly possible today (Rossiter et al. 1983). These diseases temporarily decimates some populations of wild herbivores or domestic livestock, abolishing for short periods one of the dominant forces shaping an area's vegetation (Prins and van der Jeugd 1993). Finally, a fourth important distinction is the almost universal presence of unfenced open range in Kenya and the virtual absence of such land in North America.

Despite these forces for inducing potential differences, the vegetation changes on the two continents are strikingly similar. In comparable warm regions, North American grasslands have been increasingly occupied by woody plants often of the legume family, often of the species acacia, which are spread by consumption of highly palatable fruits, and which are usually capable of fixing free nitrogen, (Hastings and Turner 1965, Archer et al. 1988). These are the same changes we have seen in Kenya.

Our findings tend to support, in large part, an elephant-giraffe-fire model, proposed to explain woodland-grassland cycles observed on the Serengeti (Pellew 1983). The photograph stations documenting large shifts from grassland to woodland represent an extreme scenario where all three major components of the model are missing. Because return of elephants, giraffes, and fire is unlikely at the sites north of Isiolo (Plates 8-10), this woodland is no longer at some temporary transitional stage and, in terms of human purpose, is locked in the woodland state.

Because our observations were confined to areas in which large herbivores were absent or at low density, we found no instances where the trend from woodland to grassland, noted elsewhere (Laws 1970, Bourlière and Hadley 1983, Dagg and Foster 1976), was evident. Elephants uprooted trees at Archer's Post, but not in numbers required to halt increased establishment of acacia (Plate 3c). Their absence may have allowed the stature of riparian trees to increase along the Athi River in Nairobi National Park (Plate 12).

One more comparison between the continent of Shantz' birth and Africa must be made. There is almost complete lack of synchroneity between fluxes in populations of humans, their domesticates, and wild herbivores in semiarid regions of the two continents. In North America, most megafaunal species have been extirpated; roughly 72 percent of the 61 species formerly present there were lost at about the time the continental glaciers were retreating. This period, at the onset of the Holocene, roughly 11,000 years before the present time (Kurtén and Anderson 1980), also marked the time of human arrival to that region (Martin and Wright 1967). Spreading across the continent and beyond, humans increasingly dominated the landscape with but few species of large mammals remaining to share the terrain. This millennia-long, mammal-depauperate condition lasted until about 500 years ago when people began introducing domesticated animals. Population peaks of these latecomers were reached only during the last century, however.

The North American status of these animal groups begs comparison with that of Africa, where they have a completely different past history but similar current status. As the cradle of human origin, the African continent supported its diverse native megafauna along with humans through much longer periods than did the North American continent (Klein 1984). In addition, domesticated herbivores began appearing in Africa several thousand years ago (Marshall 1989) but, as with North America, peak numbers of these animals were reached only during the last one hundred years.

Thus, despite the different histories of the two continents, the endpoints are remarkably similar: populations of humans and their domesticates have only recently reached all-time highs on both, and wild herbivore diversity is low or declining. In Kenya, the amazingly abundant and varied wild animal populations are gone, save for the few pockets within parks and game preserves; in North America, the ancient decimation of these animals can hardly be corrected by isolated population surges now seen among a few remnants such as elk and bison. And on both continents in areas now devoid of native megabrowsers, former grasslands are becoming increasingly dominated by woody plants.

Conclusions

Recent shifts in climate are known to have dramatic effects upon vegetation in many areas of the world. What do we see in our views of Kenyan rangelands? The impact of the sub-Sahelian drought that persisted through the 1970s and 1980s (Nicholson 1983, Kerr 1985, Lamb 1982, 1983) was felt several degrees to the north of our area, but there is little evidence of its presence this far south. Although persuasive arguments have been given for climatically induced Sahelian desertification (Bell 1987) and climatic drying undoubtedly applies over large areas of the sub-Saharan region (Nicholson 1983, Kerr 1985, Lamb 1982, 1983), monthly rainfall records for our area do not show severe protracted droughts.

Our findings are similar to those of a study of rainfall variability at Machakos, based upon a 95-year record, which found no linkage between climate and any of the environmental changes observed there (Tiffen et al. 1994). Admittedly, we may not have detected certain changes in climate, such as shifts in the frequency of high- and low-intensity rains, because a detailed analysis of daily climatic records was beyond the scope of this work. Although an increase in the amount of rain in the dry season has been shown to have a telling effect on ruminant populations in the Serengeti (Sinclair 1979), our analysis would not detect this. We recognize that brief periods of aridity have at times greatly affected livestock and wildlife mortality (Hillman and Hillman 1977) and may have serious impact on agricultural production (Downing et al. 1989, Parry 1990). Yet, even in view of these known effects, the role of climate in the changes we observed is probably minor at most. Recent worldwide reviews of shrubland expansion mechanisms support the view that the strongest link lies in the direct and indirect effects of livestock grazing, rather than factors such as historic changes in climate or atmospheric CO_2 (Archer 1994, Archer et al. 1995).

To the extent that grazing pressure is mainly by domestic livestock, this pressure will probably decline during periods of economic depression. In such times, livestock vaccination programs become prohibitively expensive and livestock mortality therefore rises from diseases no longer kept at bay. Grazing pressures on lands adjacent to national parks are driven by a unique set of conditions. "Government animals" (wildlife) move seasonally from parks onto private lands under legislative protection, whereas the privately owned livestock are not permitted to move onto the government land inside the park boundaries (Western 1975). As human population increases, the demand for additional livestock also increases without any comparable reduction in wildlife numbers. The adjacent economically important wildlife populations are promoted as tourist attractions and the seasonal incursions of these animals onto the private lands is viewed with increasing hostility (Western 1975). Recent developments, however, demonstrate the willingness of traditional Maasai pastoralists and government functionaries to settle long-standing differences over the use of the vast Amboseli region, to the mutual benefit of both parties. This gives hope that resilient human temperament can help resolve many problems in the future (Western 1994).

We wish to emphasize that although human population increase is seen as a problem at several of the camera stations, in the short term not all results of high population levels are negative. Human ingenuity and adaptability have enabled farmers to keep pace with population growth through agricultural diversification and economic flexibility (Hyden et al. 1993, Tiffen et al. 1994). It may be that only in areas such as between Meru and Isiolo, (where marginal conditions provide little promise for success among the region's smallholders), or those north of Isiolo, (where loss of grass to woody vegetation inevitably shifts the local economy away from its ancient cattle base), will serious degradation of resources occur in the short term.

But what of the long term? In 1950, western Europe contained more people than Africa. By 1993, in a brief four decades, Africa's population has risen to nearly twice that of western Europe (Brown and Kane 1994). This large population increase, the result of high birth rates and lowered mortality rates, may already have headed parts of the African continent, including Kenya, toward a "demographic trap." When population expansion reaches a point where demand exceeds the sustainable yield of local resources, a downward spiral ensues. The resulting ecological deterioration and economic decline feed on each other. As grasslands or croplands disappear, soil erosion, reduced land productivity, and failing wells follow. The resulting loss of food production and income feeds back, consuming the region's ecological base (Brown and Kane 1994).

But it would be inaccurate to totally blame burgeoning human population. Some of the forces are largely political and cannot be overcome by altering human birth rates, by improving range management practices, or by launching grand irrigation schemes. For example, much of the northern Kenyan districts of Isiolo and Turkana are uninhabited and understocked (Hogg 1987). This region, home of the Shifta Wars, has long suffered from stock raiding. This practice, partly a consequence of external forces, has restricted herd mobility and cut off access to needed dry-season pastures. For the majority of Kenya's northern arid-region inhabitants, the solution would seem to be restoration and preservation of the mobility of the pastoralists and their livestock. This is a recourse that requires unlikely action: improved security over a vast area.

The unique possibility that Homer Shantz provided for "a second or third look" illustrates the importance of maintaining archives of historical documents. Remarking about the changes Shantz recorded on film in East Africa between 1920 and 1957, Pratt and Gwynne wrote, "The results . . . are remarkable . . . for the small extent of the changes recorded" (Pratt and Gwynne 1977: p. 23). As we have found, what was certainly true after the first third of a century does not apply after passage of the second third. Without well-maintained archives, this comparison could not have been possible. The value of having records for more than two points in time has also been amply illustrated.

Our study of landscape changes is aimed primarily at describing change and secondarily at ascribing possible causes for the change. We have attempted to document in photographically unequivocal detail the status of the Kenyan landscape at two, and in many cases, three points in time. As we anticipated, our fleeting visits to the study sites have raised more questions than they have answered. Land-use history, not readily available for each site, remains the weakest link in nearly all studies of historic vegetation change, in Africa as elsewhere in the world (Bahre 1991). We have attempted, by use of these few examples, to illustrate the complexity of landscape use impacts. The list of agents directing this drama of changing landscapes includes cultural habits, politically motivated regulations, and economic reality. These forces, which might be considered aspects of human ecology, often overwhelm the role of "natural" agents such as climate or fire and make complete assessment of cause next to impossible.

Imperfect as the technique may be, repeat photography is the single best means for recording landscape features. The technique can only be enhanced, not replaced, by verbal and quantitative descriptions. Establishment in space and time of these Kenyan photographic benchmarks has provided the means for judging the rate, nature, and direction of change. Homer Leroy Shantz has left a rich photographic legacy which we and future observers can study and ponder.

Literature cited

Anonymous. 1970. National Atlas of Kenya. Survey of Kenya: Nairobi. (3d Edition).

_____. 1990. Kenya: Natural resources profile. Overseas Development Administration, Natural Resources Institute: Chatham Maritime, U.K.

_____. 1992. World population data sheet. Population Reference Bureau: Washington, D. C.

Agnew, A. D. Q. 1974. Upland Kenya wildflowers. Oxford University Press: London.

Archer, S. 1994. Woody plant encroachment into southwestern grasslands and savannas: rates, patterns and proximate causes. Pages 13-68 in M. Vavra, W. Laycock, and R. Pieper (eds.), Ecological implications of livestock herbivory in the West. Society for Range Management: Denver.

_____, D. S. Schimel, and E. A. Holland. 1995. Mechanisms of shrubland expansion: land use, climate or CO2? Climate Change 29(1):91-99.

_____, C. J. Scifres, C. R. Bassham, and R. Maggio. 1988. Autogenic succession in a subtropical savanna: rates, dynamics and processes in the conversion of a grassland to a thorn woodland. Ecological Monographs 58:111-127.

Bahre, C. J. 1991. A legacy of change. University of Arizona Press: Tucson.

Baumer, M. 1983. Notes on trees and shrubs in arid and semi-arid regions. Food and Agriculture Organization of the United Nations: Rome.

Bell, R. H. V. 1987. Conservation with a human face: conflict and reconciliation in African land use planning. Pages 79-101 in D. Anderson and R. Grove (eds.), Conservation in Africa: people, policies and practice. Cambridge University Press: Cambridge.

Belsky, A. J. 1992. Effects of grazing, competition, disturbance and fire on species composition and diversity in grassland communities. Jour. of Vegetation Science 3:187-200.

Bernard, F. E. 1972. East of Mount Kenya: Meru agriculture in transition. Humanities Press: New York, N. Y.

_____. 1993. Increasing variability in agricultural production: Meru District, Kenya, in the twentieth century. Pages 80-113 in B. L. Turner II, G. Hyden, and R. W. Kates, Population Growth and Agricultural Change in Africa. University Press of Florida: Miami.

Bogdan, A. V. 1958. A revised list of Kenya grasses. Government Printer: Nairobi.

_____, and D. J. Pratt. 1961. Common acacias of Kenya. Government Printer: Nairobi.

Bourlière, F., and M. Hadley. 1983. Present-day savannas: an overview. Pages 1-17 in F. Bourlière, Ecosystems of the World 13: Tropical Savannas. Elsevier: New York.

Brown, L. R., and H. Kane. 1994. Full house: Reassessing the earth's population carrying capacity. W. W. Norton: New York.

Campbell, K., and M. Borner. 1995. Population trends and distribution of Serengeti herbivores: implications for management. Pages 117-145 in A. R. E. Sinclair and P. Arcese (eds.), Serengeti II: dynamics, management, and conservation of an ecosystem. University of Chicago Press: Chicago.

Coe, M., and H. Beentje. 1991. A field guide to the acacias of Kenya. Oxford UniversityPress: N. Y.

Collett, D. 1987. Pastoralists and wildlife: image and reality in Kenya Maasailand. Pages 129-148 in D. Anderson and R. Grove (eds.), Conservation in Africa: people, policies and practice. Cambridge University Press: Cambridge.

Dagg, A. I., and J. B. Foster. 1976. The giraffe: its biology, behavior, and ecology. Van Nostrand Rheinhold: New York.

Dale, I. R., and P. J. Greenway. 1961. Kenya trees and shrubs. Buchanan's Kenya Estates, Ltd.: Nairobi.

Deshmukh, I. K. 1986. Primary production of a grassland in Nairobi National Park, Kenya. Journal of Applied Ecology 23:115-123.

Downing, T. E., K. W. Gitu, and C. M. Kamau. 1989. Coping with drought in Kenya. Lynne Rienner Publishers: Boulder and London.

Edwards, D. C., and A. V. Bogdan. 1951. Important grassland plants of Kenya. Pitman: Nairobi.

Glover, P. E., and M. D. Gwynne. 1961. The destruction of Maasailand. New Scientist 249:450-453.

Hastings, J. R., and R. M. Turner. 1965. The Changing Mile. University of Arizona Press: Tucson.

Heady, H. F. 1966. Influence of grazing on the composition of Themeda triandra grassland. East African Journal of Ecology 54:705-727.

Hillman, J. C., and A. K. K. Hillman. 1977. Mortality of wildlife in Nairobi National Park, during the drought of 1973-1974. East African Wildlife Journal 15:1-18.

Homewood, K. M., and W. A. Rodgers. 1991. Maasailand ecology, pastoralist development and wildlife conservation in Ngorongoro, Tanzania. Cambridge University Press: New York, N. Y.

Kemei, I. K. 1986. Vegetation-environment relationships at the National Range Research Station, Kiboko, Kenya. Pages 287-296 in R. M. Hansen, B. M. Woie, and R. D. Child (eds.), Range Development and Research in Kenya. Winrock International Institute for Agricultural Development: Morrilton, Arkansas.

Kerr, R. A. 1985. Fifteen years of African drought. Science 227:1453-1454.

Klein, R. G. 1984. Mammalian extinctions and stone age people in Africa. Pages 553-573 in P. S. Martin and R. G. Klein (eds.), Quaternary extinctions: A prehistoric revolution. University of Arizona Press: Tucson.

Kurtén, B., and E. Anderson. 1980. Pleistocene mammals of North America. Columbia University Press: New York.

Lamb, P. J. 1982. Persistence of sub-Saharan drought. Nature 299:46-48.

_____. 1983. Sub-Saharan rainfall update for 1982: continued drought. Journal of Climatology 3:419-422.

Lamprey, H. 1979. Structure and functioning of the semi-arid grazing land ecosystem of the Serengeti region (Tanzania). Pages 562-601 in UNESCO/UNEP/FAO.

Laws, R. M. 1970. Elephants as agents of habitat and landscape change in east Africa. Oikos 21:1-15.

LeHouérou, H. N. 1989. The Grazing Land Ecosystems of the African Sahel. Springer-Verlag: Berlin

Lind, E. M., and M. E. S. Morrison. 1974. East African Vegetation. Longman: London.

Lock, J. M., and T. R. Milburn. 1977. The seed biology of Themeda triandra Forsk. in relation to fire. Pages 337-349 in E. Duffy and A. J. Watt (eds.), The scientific management of animal and plant communities for conservation. Blackwell: Oxford.

Lusigi, W. J. 1984. Integrated resource assessment and management plan for Western District, Marsabit, Kenya. IPAL Technical Report Number A-6. UNESCO-FRG-MAB Integrated Project in Arid Lands.

McCarthy, F. D., and W. M. Mwangi. 1982. Kenyan Agriculture: Toward 2000. International Institute for Applied Systems Analysis: Laxenburg, Austria.

McEvedy, C., and R. Jones. 1979. Atlas of World Population History. Facts on File: New York.

McGinnies, W. J., H. L. Shantz, and W. G. McGinnies. 1991. Changes in vegetation and land use in eastern Colorado: A photographic study, 1904-1986. United States Department of Agriculture, Agricultural Research Service, ARS-85.

Martin, P. S., and H. E. Wright. 1967. Pleistocene extinctions: the search for a cause. Yale University Press: New Haven.

Marshall, F. 1989. Rethinking the role of Bos indicus in sub-Saharan Africa. Current Anthropology 30(2):235-246.

Miller, Charles. 1987. The Lunatic Express. Westlands Sundries, Ltd.: Nairobi. (2d Edition).

Miller, R. F., and P. E. Wigand. 1994. Holocene changes in semiarid pinyon-juniper woodlands: response to climate, fire, and human activities in the US Great Basin. Bioscience 44 (7):465-474.

Odingo, R. S. 1971. Settlement and rural development in Kenya. Pages 162-176 in S. H. Ominde (ed.), Studies in East African Geography and Development. University of California Press: Berkeley and Los Angeles.

Osgood, E. S. 1929. The day of the cattleman. University of Chicago Press: Chicago.

Owako, F. N. 1971. Machakos land and population problems. Pages 177-192 in S. H. Ominde (ed.), Studies in East African Geography and Development. University of California Press: Berkeley and Los Angeles.

Owen-Smith, R. N. 1988. Megaherbivores: the influence of large body size on ecology. Cambridge Studies in Ecology. Cambridge University Press: Cambridge.

Parry, M. 1990. Climate change and world agriculture. Earthscan Publications Ltd.: London.

Pellew, R. A. P. 1983. The impacts of elephant, giraffe and fire upon the Acacia tortilis woodlands of the Serengeti. African Journal of Ecology 21:41-74.

Population Reference Bureau. 1992. World Population Data Sheet

Pratt, D. J. 1967. Notes on the dispersal and germination of some tree seeds through the agency of mammals and birds. East African Wildlife Journal 5:179-180.

Pratt, D. J., and M. D. Gwynne. 1977. Rangeland Ecology and Management in East Africa. Hodder and Stoughton: London.

Pratt, D. J., and J. Knight. 1971. Bush control studies in the drier areas of Kenya. V. Effects of controlled burning and grazing management on Tarchonanthus/Acacia thicket. Journal of Applied Ecology 8:217-237.

Prins, H. H. T., and H. P. van der Jeugd. 1993. Herbivore population crashes and woodland structure in East Africa. Journal of Ecology 81:305-314.

Rogers, G. F., H. E. Malde, and R. M. Turner. 1984. Bibliography of repeat photography for evaluating landscape change. University of Utah Press: Salt Lake City.

Rose-Innes, R. 1977. A manual of Ghana grasses. Ministry of Overseas Development: Surrey, UK.

Rossiter, P. B., D. M Hassett, J. S. Wafult, L. Harstad, S. Chemwa, et al. 1983. Re-emergence of rinderpest as a threat in East Africa since 1979. Veterinary Record 113:459-61.

Saggerson, E. P. 1969. Geology. Pages 67-94, in W. T. W. Morgan (ed.), East Africa: its peoples and resources. Oxford University Press: Nairobi.

Schmutterer, H. 1979. Plants of East Africa with special reference to Kenya. Tz-Verlagsgesellschaft: Rossdorf bei Darmstadt.

Scott, R. M. 1969. Soils. Pages 95-105, in W. T. W. Morgan (ed.) East Africa: its peoples and resources. Oxford University Press: Nairobi.

Shantz, H. L. n.d. Personal Papers, Special Collections, Library, University of Arizona.

_____. 1940. Agricultural Regions of Africa. Economic Geography 16(2):122-161.

_____. and C. F. Marbut. 1923. The vegetation and soils of Africa. American Geographical Society: New York.

_____, and B. L. Turner. 1958. Photographic documentation of vegetational changes in Africa over a third of a century. University of Arizona, College of Agriculture Report 169: Tucson.

Sinclair, A. R. E. 1979. The eruption of the ruminants. Pages 82-103 in A. R. E. Sinclair and M. Norton-Griffiths (eds.), Serengeti: Dynamics of an ecosystem. University of Chicago Press: Chicago.

Sindiga, Isaac. 1984. Land and population problems in Kajiado and Narok, Kenya. African Studies Review 27(1):23-39.

Sorrenson, M. P. K. 1968. Origins of European settlement in Kenya. Oxford University Press: Nairobi.

Survey of Kenya. 1970. National Atlas of Kenya. Survey of Kenya: Nairobi.

Talbot, L. M. 1972. Ecological consequences of rangeland development in Masailand, East Africa. Pages 694-711 in M. T. Farvar and J. P. Milton (eds.), The Careless Technology: Ecology and International Development. Natural History Press: Garden City, N.Y.

Thomas, D. B., and D. J. Pratt. 1967. Brush control studies in the drier areas of Kenya. IV. Effects of controlled burning on secondary thicket in upland Acacia woodland. Journal of Applied Ecology 4:327-335.

Tiffen, M., M. Mortimore, and F. Gichuki. 1994. More people, less erosion: Environmental recovery in Kenya. John Wiley: New York, N. Y.

Tignor, R. L. 1976. The colonial transformation of Kenya.: The Kamba, Kikuyu, and Maasai from 1900 to 1939. Princeton University Press: Princeton.

Trollope, W. S. W. 1982. Ecological effects of fire in South African savannas. Pages 292-306 in B. J. Huntley and B. H. Walker (eds.), Ecology of tropical savannas. Springer-Verlag: Berlin.

Vapper, D. M. 1965. Grasses of Tanganyika. Government Printers: Dar-es-Salaam.

Ward, W. E. F., and L. W. White. 1971. East Africa: A century of change. George Allen & Unwin: London.

Wasserman, G. 1976. Politics of decolonization: Kenya Europeans and the land issue 1960-1965. Cambridge University Press: Cambridge.

Western, D. 1975. Water availability and its influence on the structure and dynamics of a savannah large mammal community. East African Wildlife Journal 13:265-286.

_____.1994. Ecosystem conservation and rural development: the case of Amboseli. Pages 15-52 in D. Western and R. M. Wright (eds.), Natural connections: Perspectives in community based conservation. Island Press: Washington, D.C.

White, F. 1983. The vegetation of Africa: A descriptive memoir to accompany the UNESCO/AETFAT/UNSO vegetation map of Africa. UNESCO.

Wilson, Alison. 1989. Samburu, Buffalo Springs and Shaba National Reserves: A visitors guide book. Friends of Conservation: Nairobi.

Young, T. P., and C. Francombe. 1991. Growth and yield estimates in natural stands of leleshwa (Tarconanthus camphoratus). Forest Ecology and Management 41:309-321.

APPENDIX

Appendix 1. Camera station name, estimated annual precipitation, elevation, latitude, and longitude for each plate number

Vegetation	Plate No.	Station Name	Annual ppn. (mm)*	Elevation (m)	Latitude	Longitude
Dwarf-shrub grassland	1	Sabachi	340	860	00°43'00" N	37°39'00" E
	2	Nakadeli	340	840	00°37'00" N	37°38'08" E
	3	Nakadeli Ranger's Post	350	840	00°37'06" N	37°38'08" E
	4	Old Archer's Post	350	840	00°37'06" N	37°38'08" E
Grassland	5	Konza Station	475	1580	00°44'27" S	37°07'42" E
	6	Kaputei Plains	500	1630	00°38'06" S	32°02'07" E
	7	Ziwani Siding	500	900	03°22'50" S	37°47'34" E
	8	Ngare Mara Plain I	550	950	00°30'17" N	37°38'32" E
	9	Ngare Mara Plain II	550	950	00°30'17" N	37°38'32" E
	10	Ngare Mara Plain III	550	950	00°30'17" N	37°38'32" E
	11	Stony Athi Station	550	1600	00°37'25" S	37°01'45" E
	12	Athi Plains	550	1600	00°22'04" S	36°56'04" E
	13	Sultan Hamud	750	900	02°00'08" S	37°23'15" E
	14	Longonot	750	2150	00°52'58" S	36°29'54" E
	15	Mbuyuni	575	1000	03°23'47" S	37°52'10" E
	16	Isiolo-Meru I	1025	1325	00°14'28" N	37°38'43" E
	17	Isiolo-Meru II	1025	1325	00°14'28" N	37°38'43" E
	18	Isiolo-Meru IIIa	1025	1325	00°14'28" N	37°38'43" E
	19	Isiolo-Meru IIIb	1025	1325	00°14'28" N	37°38'43" E
	20	Sagana	1050	1190	00°39'51" S	37°12'25" E
	21	Kimathi	1050	1235	00°40'40" S	37°11'20" E
	22	Meru	1325	1645	00°03'37" N	37°38'44" E
Bushland	23	Kijabe Ia	1025	2030	00°55'57" S	36°34'03" E
	24	Kijabe Ib	1025	2030	00°55'57" S	36°34'03" E
	25	Kijabe Ic	1025	2030	00°55'57" S	36°34'03" E
	26	Kijabe II	1025	2030	00°55'57" S	36°34'03" E
Woodland	27	Mashoti Station	575	970	03°30'10" S	38°15'05" E
	28	Bura Station	625	975	03°29'52" S	38°19'03" E
	29	Taita Hills I	1100	1220	03°27'18" S	38°21'05" E
	30	Taita hills II	1100	1220	03°27'18" S	38°21'05" E
	31	Taita Hills III	1100	1220	03°27'18" S	38°21'05" E
	32	Taita Hills IV	1100	1220	03°27'18" S	38°21'04" E
	33	Taita Hills V	1100	1220	03°27'18" S	38°21'05" E
	34	Taita Hills VIa	1250	1430	03°25'24" S	38°21'45" E
	35	Taita Hills VIb	1250	1430	03°25'24" S	38°21'45" E
	36	Taita Hills VIc	1250	1430	03°25'24" S	38°21'45 "E
	37	Runyenje Ia	1250	1450	00°25'59" S	37°33'26" E
	38	Runyenje Ib	1250	1450	00°25'59" S	37°33'26" E
	39	Runyenje Ic	1250	1450	00°25'59" S	37°33'26" E
Forest	40	Kikuyu Escarpment	1125	2160	00°56'13" S	36°35'28" E

Appendix 1. Camera station name, estimated annual precipitation, elevation, latitude, and longitude for each plate number (Continued)

Vegetation	Plate No.	Station Name	Annual ppn. (mm)*	Elevation (m)	Latitude	Longitude
Riverine Forest	41	Athi River	550	1520	00°26'57" S	36°58'22" E
	42	Thika I	800	1480	01°01'22" S	37°04'02" E
	43	Thika II	800	1480	01°01'14" S	37°04'10" E
	44	Thika III	800	1480	01°01'14" S	37°04'10" E
	45	Tana River I	1075	1190	00°39'51" S	37°12'51" E
	46	Tana River II	1075	1190	00°39'51" S	37°12'51" E
	47	Thuchi River	1275	1430	00°22'28" S	37°35'02" E
Cultivated/Disturbed	48	Mwatate	500	885	03°15'24" S	38°22'26" E
	49	Voi	550	990	03°23'39" S	38°34'03" E

*Precipitation values estimated from Survey of Kenya (1970), and rounded to closest 25 mm.

Appendix 2. Photographs showing plate no., camera station name, photograph date, negative no., and direction of view

Plate No.	Camera Station	Date Photograph Taken 1920	1924	1957	1992	Neg.# 1920	Neg.# 1924	Neg.# 1957	Neg.# 1992	True Bearing
1	Sabachi	11-Jun		22-Feb	12-Jun	38510		J-11-57	1819	N45W
2	Nakadeli	15-Jun		22-Feb	13-Jun	38548		J-4-57	1820	N80W
3	Nakadeli Ranger Post	15-Jun		22-Feb	13-Jun	38546		J-8-57	1822	N45W
4	Old Archer's Post	15-Jun		22-Feb	29-Jul	38547		J-6-57	1864	S40W
5	Konza Station	20-Apr			9-Jul	38288			1845	N25W
6	Kaputei Plains	20-Apr			9-Jul	38294			1846	N75E
7	Ziwani Siding	19-Apr			18-Jul	38263			1847	S35W
8	Ngare Mara Plain I	15-Jun			15-Jun	38557			1828	S35E
9	Ngare Mara Plain II	15-Jun			15-Jun	38556			1829	N55E
10	Ngare Mara Plain III	15-Jun			29-Jul	38552			1866	S40E
11	Stony Athi Station	20-Apr			8-Jul	38293			1844	N60E
12	Athi Plains	20-Apr			8-Jul	38296			1843	S10W
13	Sultan Hamud	20-Apr			22-Jul	38290			1861	N5E
14	Longonot	9-Jul			8-Jun	38666			1818	N10E
15	Mbuyuni	19-Apr			18-Jul	38264			1848	N17W
16	Isiolo-Meru I	27-May			14-Jun	38452			1824	N40W
17	Isiolo-Meru II	27-May			14-Jun	38460			1825	N40E
18	Isiolo-Meru IIIa	27-May			15-Jun	38455			1827a	S40W
19	Isiolo-Meru IIIb	27-May			15-Jun	38453			1827b	S60W
20	Sagana	2-May			27-Jun	38314			1835	S35E
21	Kimathi	2-May			27-Jun	38317			1838	N70W
22	Meru	30-May			17-Jun	38474			1831	S80E
23	Kijabe Ia	6-Jul		24-Jan	6-Jul	38650		G-2-57	1815a	N80W
24	Kijabe Ib	6-Jul		24-Jan	6-Jul	38649		G-1-57	1815c	N40W
25	Kijabe Ic	6-Jul		22-Jan	6-Jul	38648		F-12-57	1815b	N20E
26	Kijabe II	6-Jul			6-Jul	38646			1816	N60W
27	Mashoti Station	19-Apr			18-Jul	38268			1849	N15W
28	Bura Station	19-Apr			19-Jul	39269			1851	N60E
29	Taita Hills I		20-Feb	7-Feb	19-Jul		G-11-24	G-11-57	1852	N25E
30	Taita Hills II		21-Feb		20-Jul		H-5-24		1854	S50W
31	Taita Hills III		21-Feb		20-Jul		I-4-24		1855	S35E
32	Taita Hills IV		21-Feb		20-Jul		H-8-24		1856	S80E
33	Taita Hills V		21-Feb		20-Jul		H-6-24		1857	S0W
34	Taita Hills VIa			29-Jan	21-Jul			Ga-7-57	1858a	S70E
35	Taita Hills VIb			29-Jan	21-Jul			Ga-8-57	1858b	S45E

Appendix 2. Photographs showing plate no., camera station name, photograph date, negative no., and direction of view (Continued)

		Date Photograph Taken								
Plate No.	Camera Station	1920	1924	1957	1992	Neg.# 1920	Neg.# 1924	Neg.# 1957	Neg.# 1992	True Bearing
36	Taita Hills VIc			29-Jan	21-Jul			Ga-9-57	1858c	S15E
37	Runyenje Ia	3-May		14-Feb	26-Jun	38327		H-3-57	1832a	S80W
38	Runyenje Ib	3-May		14-Feb	28-Jul	38328		H-12-57	1862	N35W
39	Runyenje Ic	3-May		14-Feb	28-Jul	38329		H-6,7-57	1863	N3E
40	Kikuyu Escarpment	5-Jul		22-Jan	8-Jun	38634		F-11-57	1817	N45E
41	Athi River	20-Apr			8-Jul	38295			1842	S80E
42	Thika I	2-May		17-Feb	28-Jun	38304		J-2-57	1839	S30W
43	Thika II	18-May		24-Jan	28-Jun	38405		G-4-57	1840	S40W
44	Thika III	19-May		24-Jan	28-Jun	38406		G-8-57	1841	S70E
45	Tana River I	2-May			27-Jun	38312			1836	N5W
46	Tana River II	2-May			27-Jun	38313			1837	S0W
47	Thuchi River	3-May		15-Feb	26-Jun	38335		I-2-57	1833	S90W
48	Mwatate	19-Apr			19-Jul	38273			1850	S15E
49	Voi	19-Apr		28-Jan	21-Jul	38276		G-10-57	1860	S10E

Appendix 3. Plant species observed at camera stations

Plate Number

	Plant name	1	2	3	4	5	6	7	8	9	10	11	12	13	14	15	16	17	18-19	20	21	22	23-25	26	27	28	29	30	31	32	33	34-36	37-39	40	41	42	43	44	45	46	47	48	49
Trees	*Acacia brevispica* Harms												X																														
	Acacia drepanolobium Harms & Sjöstedt					X	X					X	X	X	X																												
	Acacia hockii De Wild.					X														X	X							X	X		X				X								
	Acacia mellifera (Vahl) Benth.													X		X																X			X								X
	Acacia nilotica (L.) Del.					X							X	X							X				X	X																	X
	Acacia nubica Benth.								X		X																																X
	**Acacia podalyriifolia* A. Cunn.																																X										
	Acacia reficiens Wawra & Peyr.		X	X	X				X		X																																
	Acacia senegal (L.) Willd.	X						X						X		X									X							X											X
	Acacia seyal Del.												X	X							X												X			X							
	Acacia sieberiana DC.													X																													
	Acacia sp.																																X										
	Acacia tortilis (Forsic.) Hayne	X	X	X	X			X	X		X			X		X	X	X	X						X	X																X	X
	Acacia xanthophloea Benth.												X		X																				X							X	
	Acacia zanzibarica (S. Moore) Taub.										X																																
	Adansonia digitata Linn.															X																										X	X
	Adina microcephala (Del.) Hiern																																					X					

Appendix 3. Plant species observed at camera stations—Continued

Plant name	Plate Number 1	2	3	4	5	6	7	8	9	10	11	12	13	14	15	16	17	18-19	20	21	22	23-25	26	27	28	29	30	31	32	33	34-36	37-39	40	41	42	43	44	45	46	47	48	49
Albizia amara (Roxb.) Boiv.																														X		X										
Albizia gummifera (J. F. Gmel.) C. A. Sm.																																			X							
Albizia sp.																																								X		
**Araucaria columnaris* Hook.																										X					X											
**Azadirachta indica* A. Juss.																									X																	X
Balanites aegyptiaca (L.) Del.					X	X	X	X	X	X	X	X	X			X	X							X																		
Boscia angustifolia A. Rich.																X																										X
Boscia salicifolia Oliv.							X								X																											
Breonadia salicina (Vahl) Hepper & J. R. Wood																																					X					
Breonadia sp.																																								X		
Bridelia micrantha Baill.																						X	X			X			X	X		X						X		X		
Cadaba farinosa Forsk.									X																																	
Calodendrum capense (L. f.) Thunb.																																	X									
Capparis tomentosa Lam.																									X																	
**Cassia siamea* Lam.																									X																	X
**Cassia spectabilis* DC.																				X																						
**Casuarina equisetifolia* L.																									X							X										

Appendix 3. Plant species observed at camera stations—Continued

	Plate Number																																								
Plant name	1	2	3	4	5	6	7	8	9	10	11	12	13	14	15	16	17	18-19	20	21	22	23-25	26	27	28	29	30	31	32	33	34-36	37-39	40	41	42	43	44	45	46	47	48	49
Combretum molle R. Br. ex G. Don													X																													
Combretum sp.					X														X																							
Commiphora africana (A. Rich.) Engl.															X									X																		X
Commiphora baluensis Engl.																																										X
Commiphora candidula Sprague		X		X																																						
Commiphora riparia Engl.																																										X
Commiphora schimperi (Berg.) Engl.																																										X
Commiphora sp.																															X											
Cordia abyssinica R. Br.																					X											X								X		
Cordia ovalis R. Br.							X																																			
Croton macrostachyus Hochst. ex Del.																					X								X		X	X				X				X		
Croton megalocarpus Hutch.																				X						X			X			X										X
Croton sp. nr. *c. sylvaticus* Hochst.																									X																	
**Cupressus lusitanica* Mill.																						X									X											
**Cupressus* sp.																					X						X	X		X		X										
Erythrina abyssinica Lam. ex DC.																			X	X						X	X	X	X	X	X							X				
**Eucalyptus camaldulensis* Dehnh.																									X							X										

Appendix 3. Plant species observed at camera stations—Continued

	Plate Number																																								
Plant name	1	2	3	4	5	6	7	8	9	10	11	12	13	14	15	16	17	18 - 19	20	21	22	23 - 25	26	27	28	29	30	31	32	33	34 - 36	37 - 39	40	41	42	43	44	45	46	47	48	49
**Eucalyptus globulus* Labill.														X								X					X	X		X	X											X
**Eucalyptus* sp.																			X						X	X			X					X	X					X		
Euphorbia candelabrum Trem. ex Kotschy																						X			X																	
Euphorbia nyikae Pax																						X																X	X			
Euphorbia tirucalli L.																			X	X																		X				
Ficus capensis Thunb.																				X											X							X	X			X
Ficus natalensis Hochst.																											X			X	X					X		X	X			
Ficus thonningii Blume																										X										X			X			
Ficus vallis-choudae Del.																																			X	X						
Gnidia latifolia (Oliv.) Gilg.																			X																							
**Grevillea robusta* A. Cunn.					X									X						X	X				X	X	X	X	X	X	X	X						X	X	X		
Grewia trichocarpa Hochst.																														X	X	X										
**Jacaranda mimosifolia* D. Don																										X			X	X												
Juniperus procera Hochst. ex Endl.																						X									X		X									
Lannea stuhlmannii Endl.																								X	X																X	
**Leucaena leucocephala* (Lam.) deWit																									X																	
**Mangifera indica* L.																										X	X	X	X												X	X
Manilakara sp.																																					X					

Appendix 3. Plant species observed at camera stations—Continued

		Plate Number																																							
	Plant name	1	2	3	4	5	6	7	8	9	10	11	12	13	14	15	16	17	18-19	20	21	22	23-25	26	27	28	29	30	31	32	33	34-36	37-39	40	41	42	43	44	45	46	47	48	49
	Markhamia hildebrandtii (Baker) Sprague																																			X	X						
	Markhamia platycalyx Sprague																																X										
	Melia volkensii Gürke																X											X															
	Newtonia buchananii (Baker) Gilbert & Boutique																																				X	X					
	Olea africana Mill.																																	X			X						
	Olea hochstetteri Baker																																	X									
	Olea sp.																																X										
	Phoenix reclinata Jacq.																															X											
	Piptadenia sp.																																								X		
	**Psidium guajava* L.																									X	X	X	X	X												
	**Schinus molle* L.					X									X								X												X								
	Spathodea nilotica Seem.																																X										
	Sterculia rhynchocarpa K. Schum.																																									X	
	**Tamarindus indica* L.																																									X	X
	Teclea sp.																																							X			
	Turraea robusta Gürke																											X															
Shrubs	*Abutilon mauritianum* (Jacq.) Medic.																																	X									
	Acalypha fruticosa Forsk.												X										X									X					X						X

Appendix 3. Plant species observed at camera stations—Continued

	Plate Number																																								
Plant name	1	2	3	4	5	6	7	8	9	10	11	12	13	14	15	16	17	18-19	20	21	22	23-25	26	27	28	29	30	31	32	33	34-36	37-39	40	41	42	43	44	45	46	47	48	49
Acokanthera longiflora Stapf																																	X									
**Agave* sp.																																			X							
Cadaba farinosa Forsk.												X																														
Caesalpinia decapetala (Roth) Alston																															X											
Capparis tomentosa Lam.																																		X								
Cassia didymobotrya Fres.																			X																							
Cissus rotundifolia Vahl.																																						X	X			
Cordia gharaf (Fprsk.) Ehrenb. & Aschers.							X																	X																		
Cordia sinensis Lam.																																									X	
Croton dichogamus Pax																																			X							
Dodonaea viscosa (L.) Jacq.																											X				X											
Dombeya faucicola K. Schum.																																	X		X							
Duosperma eremophilum (Milne-Redhead) Brummitt		X		X																																						
Euclea divinorum Hiern																																	X									
Farsetia somalensis Engl.		X																																								
Grewia bicolor Juss.							X									X																									X	
Grewia hexamita Burret																								X																		

Appendix 3. Plant species observed at camera stations—Continued

	Plate Number																																									
Plant name	1	2	3	4	5	6	7	8	9	10	11	12	13	14	15	16	17	18-19	20	21	22	23-25	26	27	28	29	30	31	32	33	34-36	37-39	40	41	42	43	44	45	46	47	48	49
Grewia similis K. Schum.												X				X	X																	X								
Grewia tenax (Forsk.) Fiori							X	X																																		
Grewia sp.																																		X								
Hibiscus fuscus Garcke & Hutch.																															X		X									
Hibiscus aponeurus Sprague						X					X																															
Hoslundia opposita Vahl							X																																			
Ipomoea cicatricosa Baker		X		X																																						
Ipomoea sp.		X	X																																							
Kedrostis gijef (Gilg.) C. Jeffrey		X																																								
**Lantana trifolia* L.																						X					X	X	X		X											
**Lantana camara* L.																			X	X									X		X	X			X	X		X		X	X	
Leucas calostachys Oliver																				X																						
**Lippia javanica* Spreng.																						X					X					X										
Lippia ukambensis Vatke													X																													
Maytenus putterlickioides (Loes.) Exell & Mendonça												X				X								X																		
**Opuntia ficus-indica* Mill.																																		X	X							
Ormocarpum trychocarpum (Taub.) Harms.												X																														
Phytolacca dodecandra L'Harit																						X																				

Appendix 3. Plant species observed at camera stations—Continued

		Plate Number																																								
	Plant name	1	2	3	4	5	6	7	8	9	10	11	12	13	14	15	16	17	18-19	20	21	22	23-25	26	27	28	29	30	31	32	33	34-36	37-39	40	41	42	43	44	45	46	47	48	49
	Psiadia arabica Jaub. & Spach																						X					X				X											
	Ricinus communis L.																			X															X								
	Rhus vulgaris Meikle																X										X	X		X		X											X
	Rhus natalensis Bernh.													X						X			X					X	X			X					X						X
	Rubus sp.																															X											
	Salsola dendroides Pall.			X																																							
	Sericocomopsis hildebrandtii Schinz	X	X		X				X	X																																	
	Tarchonanthus camphoratus L.														X								X	X																			
	**Tithonia diversifolia* A. Gray																					X					X	X	X	X	X		X				X			X		X	
	Triumfetta rhomboidea Jacq.																			X													X										
	Volkensinia prostrata (Volkens) Schinz	X	X	X	X																																						
Forbs	*Acanthospermum hispidum* DC.																																X										
	Achyranthes aspera L.						X							X								X																		X			X
	Aerva lanata (L.) Juss.						X																																				
	Aerva persica (Burn.) Merril															X																											
	Ageratum conyzoides L.																					X								X													
	Aloe dawei Berger		X	X																																X							
	Asparagus racemosus Willd.		X																																								

Appendix 3. Plant species observed at camera stations—Continued

	Plate Number																																									
Plant name	1	2	3	4	5	6	7	8	9	10	11	12	13	14	15	16	17	18-19	20	21	22	23-25	26	27	28	29	30	31	32	33	34-36	37-39	40	41	42	43	44	45	46	47	48	49
Aspilia mossambicensis Oliv.						X							X	X						X									X	X				X				X				
Aspilia pluriseta Schweinf. ex Engl.						X					X																															
Barleria acanthoides Vahl.	X	X	X					X	X							X																										
Barleria eranthemoides R. Br.			X																																							
Basella alba L.																																						X				
Bidens pilosa L.					X																					X						X										
Bidens sp.																																								X		
Blepharis linariifolia Pers.	X	X	X																																							
Blotiela stipitata Alston																															X											
Caralluma retrospeciens N. E. Br.				X																																						
Cassia occidentalis L.															X				X	X																						
Cleome sp.																					X																					
Conyza floribunda HBK																																X										
Conyza newii Oliv. & Hiern.																														X	X											
**Cotyledon barbeyi* Schweinf. ex Penzig																																			X							
Crotalaria agatiflora Schweinf.															X																											
Cucumis dipsaceus Spach																																		X								
Cythula polycephala Moq.																					X																					

Appendix 3. Plant species observed at camera stations—Continued

Plant name	Plate Number																																								
	1	2	3	4	5	6	7	8	9	10	11	12	13	14	15	16	17	18-19	20	21	22	23-25	26	27	28	29	30	31	32	33	34-36	37-39	40	41	42	43	44	45	46	47	48	49
**Datura stramonium* L.					X																																					
Duosperma eremophyllum (Miln.-Redh.) Napper	X																																									
Euphorbia heterochroma Pax			X																																							
Galinsoga parviflora Cav.																					X					X																
Gomphocarpus semilunatus A. Rich.						X																																				
Heliotropium sp.	X																																								X	
Heliotropium steudneri Vatke															X																											
Hermania alhiensis K. Schum.							X																																			
Hibiscus vitifolius L.																			X																							
Hygrophila auriculata (Schumach.) Heine																			X																							
Hypoestes triflora (Forsk.) Roem. & Schult.																																										X
Hypoestis sp.											X																															
Impatiens sp.																																								X		
Indigofera schimperi Jaub. & Spach																								X																		
Indigofera spinosa Forsk.	X	X	X																																							
Ipomoea donaldsonii Rendle		X																																								
Ipomoea sp.	X																																								X	

Appendix 3. Plant species observed at camera stations—Continued

Plant name	Plate Number																																									
	1	2	3	4	5	6	7	8	9	10	11	12	13	14	15	16	17	18-19	20	21	22	23-25	26	27	28	29	30	31	32	33	34-36	37-39	40	41	42	43	44	45	46	47	48	49
Ipomoea wightii (Wall.) Choisy																					X										X		X									
Jasminum dichotomum Vahl																											X															
Justicia flava Vahl																																										X
Leucas calostachyus Oliv.																				X							X															
Malva sp.																																		X								
Mariscus macropus Kunth																				X																						
Melanthera scandens (Schumach. & Thonn.) Roberts																																								X		
Monechma debile Forsk.											X																														X	
Nephrolepus undulata (Afz. ex Sw.) J. Smith																													X													
Ocimum basilicum L.																											X		X	X	X											
Ocimum kilimandschari-cum Guerke																						X											X									
Ocimum laniifolium Benth.																						X																				
**Passiflora edulis* Sims																																X										
Plectranthus barbatus Ardr.																															X									X		
Pteridium aquilinum (L.) Kunth.																													X													
Sansevieria ehrenbergii Schweinf. ex Baker		X																																								

Appendix 3. Plant species observed at camera stations—Continued

	Plant name	Plate Number																																							
		1	2	3	4	5	6	7	8	9	10	11	12	13	14	15	16	17	18-19	20	21	22	23-25	26	27	28	29	30	31	32	33	34-36	37-39	40	41	42	43	44	45	46	47	48	49
	Sarcostemma adongensis L.																									X																	
	Seddera hirsuta Hall. f.	X	X	X																																							
	Sida cuneifolia Roxb.																			X																							
	Solanum dubium Fres.								X	X																																	
	Solanum incanum L.																X			X			X	X								X	X		X							X	
	Sphaeranthus suaveolens (Forsk.) DC.																			X																							
	**Tagetes minuta* L.					X																X					X						X										
	Tribulus terrestris L.																X		X																								
	**Trichodesma zeylanicum* (L.) R. Br.																																										X
	Trifolium semipilosum Fres.																					X																					
	Triumfetta rhomboidea Jacq.																				X																						
	Vernonia cinerascens Sch. Bip.																											X		X	X	X											
	Vernonia lassiopus O. Hoffm.																										X			X													
Grasses and Sedges	*Aristida adoensis* Hochst.																						X										X										
	Aristida adscensionis L.		X		X																																						
	Aristida keniensis Henr.	X	X			X										X																										X	X

Appendix 3. Plant species observed at camera stations—Continued

Plant name	1	2	3	4	5	6	7	8	9	10	11	12	13	14	15	16	17	18-19	20	21	22	23-25	26	27	28	29	30	31	32	33	34-36	37-39	40	41	42	43	44	45	46	47	48	49
Arisitida mutabilis Trin. & Rupr.		X	X						X																																	
Aristida sp.	X																																									
Bothriochloa insculpta A. Rich.					X	X						X	X			X	X		X	X																						
Brachiaria eruciformis (Sw.) Gris																										X																
Brachiaria soluta Stapf																														X												
Cenchrus setigerus Vahl												X																						X								
Cenchrus ciliaris L.							X	X	X		X	X	X		X									X										X								X
Chloris pycnothrix Trin.																			X										X													
Chloris roxburghiana Schult.							X																	X																		X
Chloris sp.		X																																								
Chrysopogon aucheri Stapf.			X						X	X																																
Cymbopogon validus Stapf ex Burtt-Davy												X																					X									
Cynodon dactylon (L.) Pers.					X	X													X	X		X							X	X		X	X	X								
Cynodon plectostachyum (Schum.) Pilger							X							X	X	X						X	X	X	X									X								X
Cyperus articulatus L.																			X	X																		X	X			
Dactyloctenium aegyptium (L.) Beauv.		X			X																																					
Dactyloctenium scindicum Boiss.																								X																		

Appendix 3. Plant species observed at camera stations—Continued

	Plate Number																																									
Plant name	1	2	3	4	5	6	7	8	9	10	11	12	13	14	15	16	17	18 - 19	20	21	22	23 - 25	26	27	28	29	30	31	32	33	34 - 36	37 - 39	40	41	42	43	44	45	46	47	48	49
Digitaria diagonalis (Nees) Stapf																					X					X				X	X	X										
Digitaria milanjiana (Rendle) Stapf							X																																			
Digitaria scalarum Chiov.					X																X	X				X						X										
Echinochloa colona (L.) Link																			X	X																						
Echinochloa pyramidalis (Lam.) Hitch. & Chase																																							X			
Enteropogon macrostachyus (H. Rich.) Benth.		X	X				X																																		X	X
Eragrostis braunii Schweinf.																						X																				
Eragrostis cilianensis (All.) Lutati		X	X	X			X								X																											
Eragrostis superba Peyr.							X																																		X	
Eragrostis tenuifolia (A. Rich.)Steud.																				X		X				X	X				X	X	X									
Harpachne schimperi A. Rich.																						X																				
Heteropogon contortus (L.) Roem. & Schult.												X												X			X														X	
Hyparrhenia collina (Pilg.) Stapf												X																														
Hyparrhenia cymbaria (L.) Stapf												X																					X									

Appendix 3. Plant species observed at camera stations—Continued

	Plate Number																																								
Plant name	1	2	3	4	5	6	7	8	9	10	11	12	13	14	15	16	17	18-19	20	21	22	23-25	26	27	28	29	30	31	32	33	34-36	37-39	40	41	42	43	44	45	46	47	48	49
Hyparrhenia hirta (L.) Stapf																		X																								
Hyparrhenia rufa (Nees) Stapf																											X															
Ischaemum afrum (J. F. Gmel.) Dandy						X					X																															
Latipes senegalensis Kunth.	X	X	X	X			X								X																											
Leptochloa obtusiflora Hochst.							X								X										X																	
Lintonia nutans Stapf						X							X																													
Oropetium capense Stapf	X																																									
Panicum maximum Jacq.						X						X	X		X				X	X				X	X	X				X	X		X	X							X	X
Paspalum scrobiculatum L.																											X					X										
Pennisetum clandestinum Chiov.					X																	X	X			X	X		X	X	X	X										
Pennisetum mezianum Leeke								X	X	X	X	X	X	X		X	X	X			X																					
Pennisetum purpureum Schumach.																					X					X	X				X	X		X						X		
Pennisetum stramineum Peter					X	X		X	X	X	X	X				X	X	X																X								
Phragmites mauritianus Kunth.																																						X	X	X		
Rynchelytrum repens (Willd.) C. E. Hubb.																																X										
Setaria sphacelata Stapf & C. E. Hubb.												X																														

Appendix 3. Plant species observed at camera stations—Continued

	Plate Number																																								
Plant name	1	2	3	4	5	6	7	8	9	10	11	12	13	14	15	16	17	18-19	20	21	22	23-25	26	27	28	29	30	31	32	33	34-36	37-39	40	41	42	43	44	45	46	47	48	49
Setaria verticillata (L.) Beauv.					X																																					
Sorghum verticillifolium (Steud.) Stapf																										X																
Sporobolus fillipes Napper																										X																
Sporobolus fimbriatus Nees							X																	X																		
Sporobolus sp.	X																																									
Sporobolus pyramidalis Beauv.																			X	X						X	X		X			X										
Tetrapogon spatheceus Desf.		X	X				X		X																																	X
Tetrapogon tenellus (Roxb.) Chiov.							X																																			
Themeda triandra Forsk.					X	X	X				X	X				X						X		X																		

* Introduced species.

Appendix 4. Scientific equivalents for common plant names used in the text and captions

acalypha	*Acalypha fruticosa*
adina	*Adina microcephala*
adrue	*Cyperus articulatus*
African fox tail	*Cenchrus ciliaris*
African pencil cedar	*Juniperus procera*
aloe	*Aloe dawei*
antelope grass	*Echinochloa pyramidalis*
bamboo grass	*Pennisetum mezianum*
baobab	*Adansonia digitata*
barkcloth fig	*Ficus natalensis*
Bermuda grass	*Cynodon dactylon*
bitter albizia	*Albizia amara*
bracken	*Pteridium aquilinum*
bridelia	*Bridelia micrantha*
broad-leaved croton	*Croton macrostachyus*
buffel grass	*Cenchrus ciliaris*
bur-marigold	*Bidens pilosa*
caltrops	*Tribulus terrestris*
candelabra euphorbia	*Euphorbia candelabrum*
Cape fig	*Ficus capensis*
Cape chestnut	*Calodendrum capense*
cassava	*Manihot esculenta*
castor oil plant	*Ricinus communis*
casuarina	*Casuarina equisetifolia*
Cook's araucaria	*Araucaria columnaris*
cotyledon	*Cotyledon barbeyi*
crab grass	*Digitaria scalarum*
croton	*Croton megalocarpus*
curse of India	*Lantana camara*
cypress	*Cupressus (unidentified species)*
desert date	*Balanites aegyptiaca*
dombeya	*Dombeya faucicola*
dropseed	*Sporobolus (unidentified species)*
Egyptian thorn	*Acacia nilotica*
Ehrenberg wild sisal	*Sansevieria ehrenbergii*
elephant grass	*Pennisetum purpureum*
euclea	*Euclea divinorum euclea*
false umbrella thorn	*Acacia reficiens*
feather fingergrass	*Chloris virgata*
fever tree	*Acacia xanthophloea*
finger euphorbia	*Euphorbia tirucalli*
flame tree	*Spathodea nilotica*
golden wattle	*Acacia podalyriifolio*

Appendix 4. Scientific equivalents for common plant names used in the text and captions—Continued

guava	*Psidium guajava*
Guinea grass	*Panicum maximum*
hairy herringbone grass	*Tetrapogon spathecius*
hook grass	*Latipes senegalensis*
hook thorn	*Acacia mellifera*
Indian fig	*Opuntia ficus-indica*
jacaranda	*Jacaranda mimosifolia*
jungle rice	*Echinochloa colona*
Kenya needlegrass	*Aristida keniensis*
Kikuyu grass	*Pennisetum clandestinum*
Kodo millet	*Paspalum scrobiculatum*
lantana	*Lantana trifolia*
large-leaved cordia	*Cordia abyssinica*
leleshwa	*Tarchonanthus camphoratus*
Maasai grass	*Pennisetum stramineum*
mango	*Mangifera indica*
manilakara	*Manilakara sp.*
markhamia	*Markhamia hildebrandtii*
melia	*Melia volkensii*
Mexican cypress	*Cupressus lusitanica*
Mexican marigold	*Tagetes minuta*
Mexican sunflower	*Tithonia diversifolia*
Naivasha star grass	*Cynodon plectostachyus*
neem tree	*Azadirachta indica*
newtonia	*Newtonia buchananii*
Nyika euphorbia	*Euphorbia nyikae*
orange-leaved croton	*Croton dichogamus*
passion fruit	*Passiflora edulis*
peacock flower	*Albizia gummifera*
pepper tree	*Schimus molle*
pyramid dropseed	*Sporobolus pyramidalis*
red oat grass	*Themeda triandra*
Red River gum	*Eucalyptus camaldulensis*
red-hot poker tree	*Erythrina abyssinica*
rhus	*Rhus natalensis*
sand olive	*Dodonaea angustifolia*
sandpaper cordia	*Cordia ovalis*
Senegal palm	*Phoenix reclinata*
silky oak	*Grevillea robusta*
sisal	*Agave sisalana*
six-weeks threeawn	*Aristida adscensionis*
Sodom apple	*Solanum incanum*
stinkgrass	*Eragrostis cilianensis*
strangler fig	*Ficus thonningii*
sweet pitted grass	*Bothriochloa insculpta*

Appendix 4. Scientific equivalents for common plant names used in the text and captions—Continued

tamarind	*Tamarindus indica*
tanglehead	*Heteropogon contortus*
Tasmanian blue gum	*Eucalyptus globulus*
three-awn	*Aristida (unidentified species)*
three-thorned acacia	*Acacia senegal*
triumfetta	*Triumfetta rhomboidea*
umbrella thorn	*Acacia tortilis*
velvet-leaved combretum	*Combretum molle*
whistling thorn	*Acacia drepanolobium*
white thorn	*Acacia hockii*
white-galled acacia	*Acacia seyal*
wild olive	*Olea africana*
wiry lovegrass	*Eragrostis tenuifolia*

Appendix 5. Common-name equivalents used in the text and captions for scientific plant name

Acacia drepanolobium	whistling thorn
Acacia hockii	white thorn
Acacia mellifera	hook thorn
Acacia nilotica	Egyptian thorn
Acacia podalyriifolia	golden wattle
Acacia reficiens	false umbrella thorn
Acacia senegal	three-thorned acacia
Acacia seyal	white-galled acacia
Acacia tortilis	umbrella thorn
Acacia xanthophloea	fever tree
Acalypha fruticosa	acalypha
Adansonia digitata	baobab
Adina microcephala	adina
Agave sisalana	sisal
Albizia amara	bitter albizia
Albizia gummifera	peacock flower
Aloe dawei	aloe
Araucaria columnaris	Cook's araucaria
Aristida adscensionis	six-weeks three-awn
Aristida keniensis	Kenya needlegrass
Azadirachta indica	neem tree
Balanites aegyptiaca	desert date
Bidens pilosa	bur-marigold
Bothriochloa insculpta	sweet pitted grass
Bridelia micrantha	bridelia
Calodendrum capense	Cape chestnut
Casuarina equisetifolia	casuarina
Cenchrus ciliaris	African fox tail, buffel grass
Chloris virgata	feather fingergrass
Combretum molle	velvet-leaved combretum
Cordia abyssinica	large-leaved cordia
Cordia ovalis	sandpaper cordia
Cotyledon barbeyi	cotyledon
Croton dichogamus	orange-leaved croton
Croton macrostachyus	broad-leaved croton
Croton megalocarpus	croton
Cupressus lusitanica	Mexican cypress
Cupressus (unidentified species)	cypress
Cynodon dactylon	common star grass, Bermuda grass
Cynodon plectostachyus	Naivasha star grass
Cyperus articulatus	adrue
Digitaria scalarum	crab grass
Dodonaea angustifolia	sand olive
Dombeya faucicola	dombeya

Appendix 5. Common-name equivalents used in the text and captions for scientific plant name—Continued

Echinochloa colona	jungle rice
Echinochloa pyramidalis	antelope grass
Eragrostis cilianensis	stinkgrass
Eragrostis tenuifolia	wiry lovegrass
Erythrina abyssinica	red-hot poker tree
Eucalyptus camaldulensis	Red River gum
Euclea divinorum	euclea
Euphorbia candelabrum	candelabra euphorbia
Euphorbia nyikae	Nyika euphorbia
Euphorbia tirucalli	finger euphorbia
Ficus capensis	Cape fig
Ficus natalensis	barkcloth fig
Ficus thonningii	strangler fig
Grevillea robusta	silky oak
Heteropogon contortus	tanglehead
Jacaranda mimosifolia	jacaranda
Juniperus procera	African pencil cedar
Lantana camara	curse of India
Lantana trifolia	lantana
Latipes senegalensis	hook grass
Mangifera indica	mango
Manihot esculenta	cassava
Manilakara sp.	manilakara
Markhamia hildebrandtii	markhamia
Melia volkensii	melia
Newtonia buchananii	newtonia
Olea africana	wild olive
Opuntia ficus-indica	Indian fig
Panicum maximum	Guinea grass
Paspalum scrobiculatum	Kodo millet
Passiflora edulis	passion fruit
Pennisetum clandestinum	Kikuyu grass
Pennisetum mezianum	bamboo grass
Pennisetum purpureum	elephant grass
Pennisetum stramineum	Maasai grass
Phoenix reclinata	Senegal palm
Psidium guajava	guava
Pteridium aquilinum	bracken
Rhus natalensis	rhus
Ricinus communis	castor oil plant
Sansevieria ehrenbergii	Ehrenberg wild sisal
Schimus molle	pepper tree
Solanum incanum	Sodom apple
Spathodea nilotica	flame tree
Sporobolus pyramidalis	pyramid dropseed

Appendix 5. Common-name equivalents used in the text and captions for scientific plant name—Continued

Tagetes minuta	Mexican marigold
Tamarindus indica	tamarind
Tarchonanthus camphoratus	leleshwa
Tetrapogon spathecius	hairy herringbone grass
Themeda triandra	red oat grass
Tithonia diversifolia	Mexican sunflower
Tribulus terrestris	caltrops
Triumfetta rhomboidea	triumfetta
Eucalyptus globulus	Tasmanian blue gum

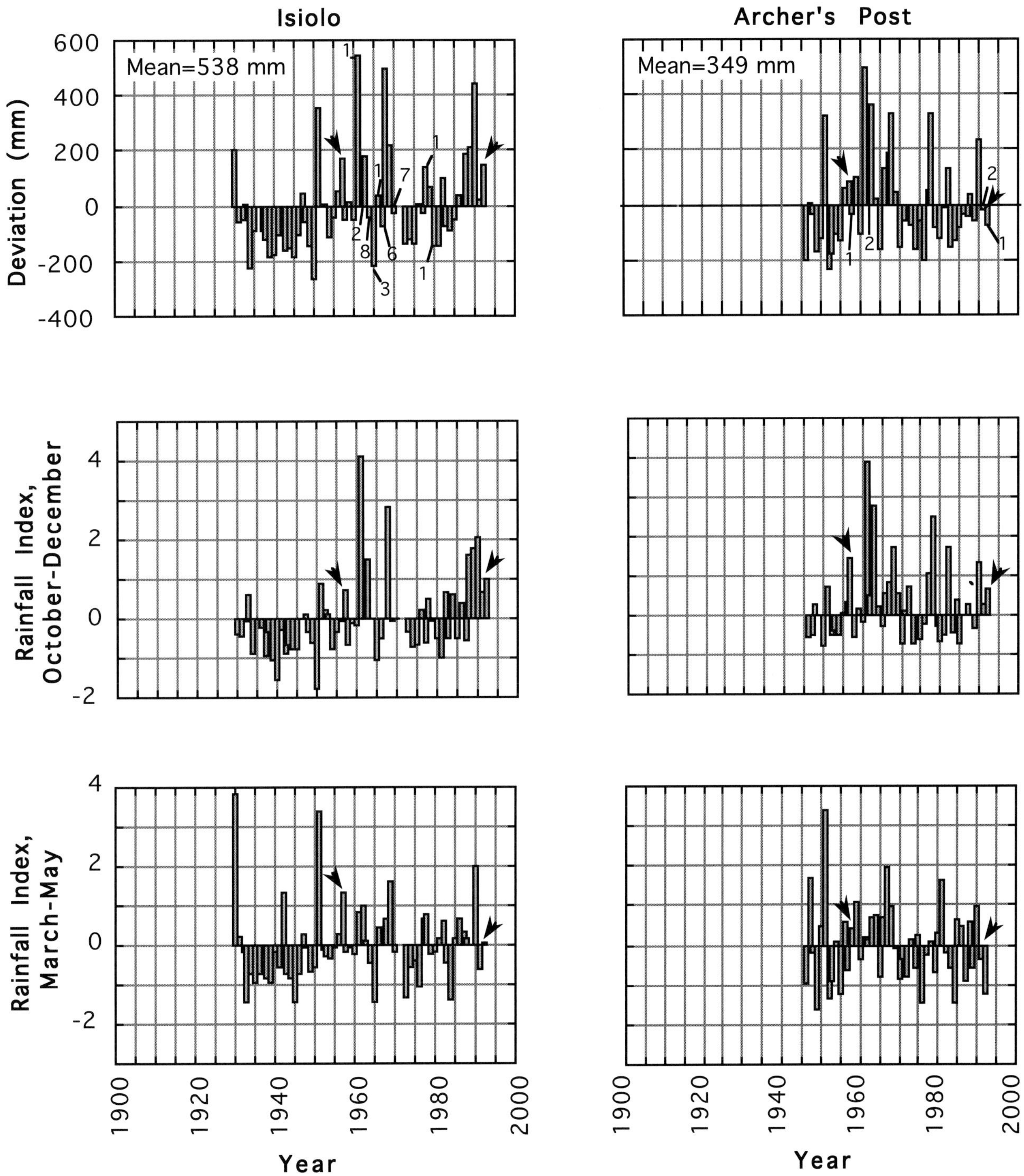

Appendix 6. Rainfall data for selected stations in Kenya. Data for each station are shown as follows: (top) chart showing yearly deviations from long-term station average; (center) chart showing yearly precipitation index (see text) for period of "short rains" (October-December); (bottom) chart showing yearly precipitation index for period of "long rains" (March-May). Arrows mark years when photographs were taken. Numerals indicate number of months for which data are missing in given years.

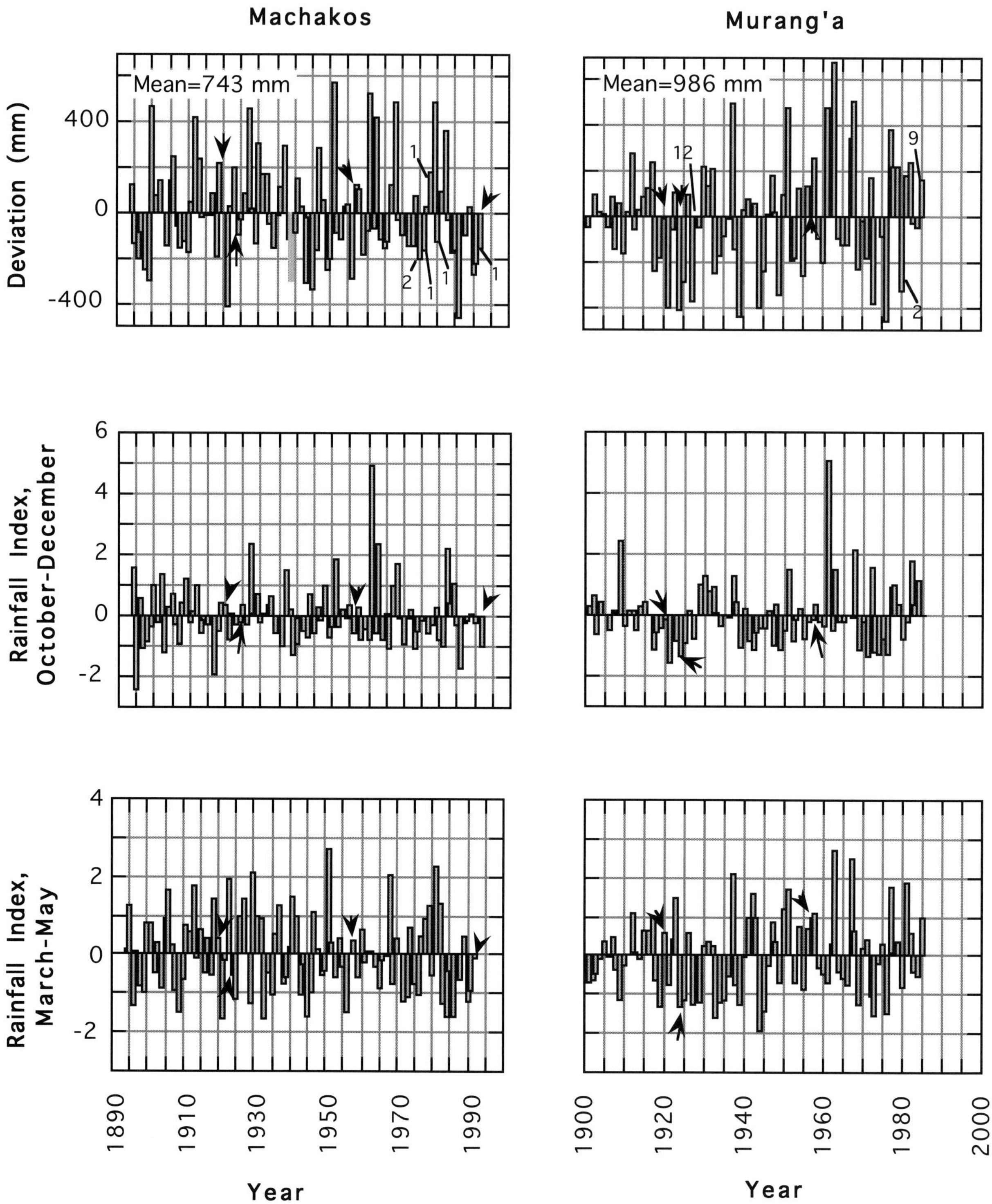
Machakos
Murang'a
Mean=743 mm
Mean=986 mm
Deviation (mm)
Rainfall Index, October-December
Rainfall Index, March-May
Year
Year

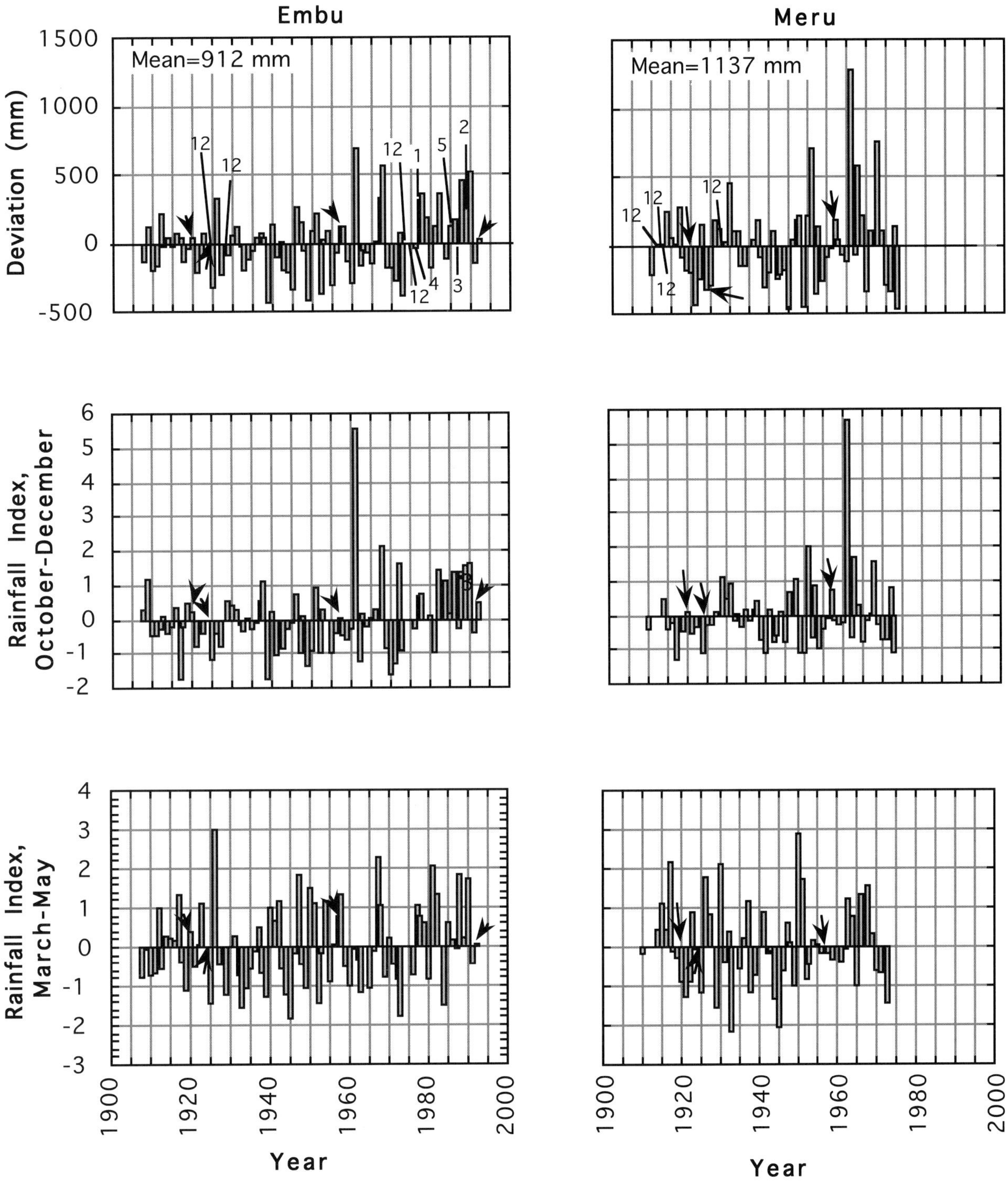
Embu
Meru
Mean=912 mm
Mean=1137 mm
Deviation (mm)
Rainfall Index, October-December
Rainfall Index, March-May
Year
Year

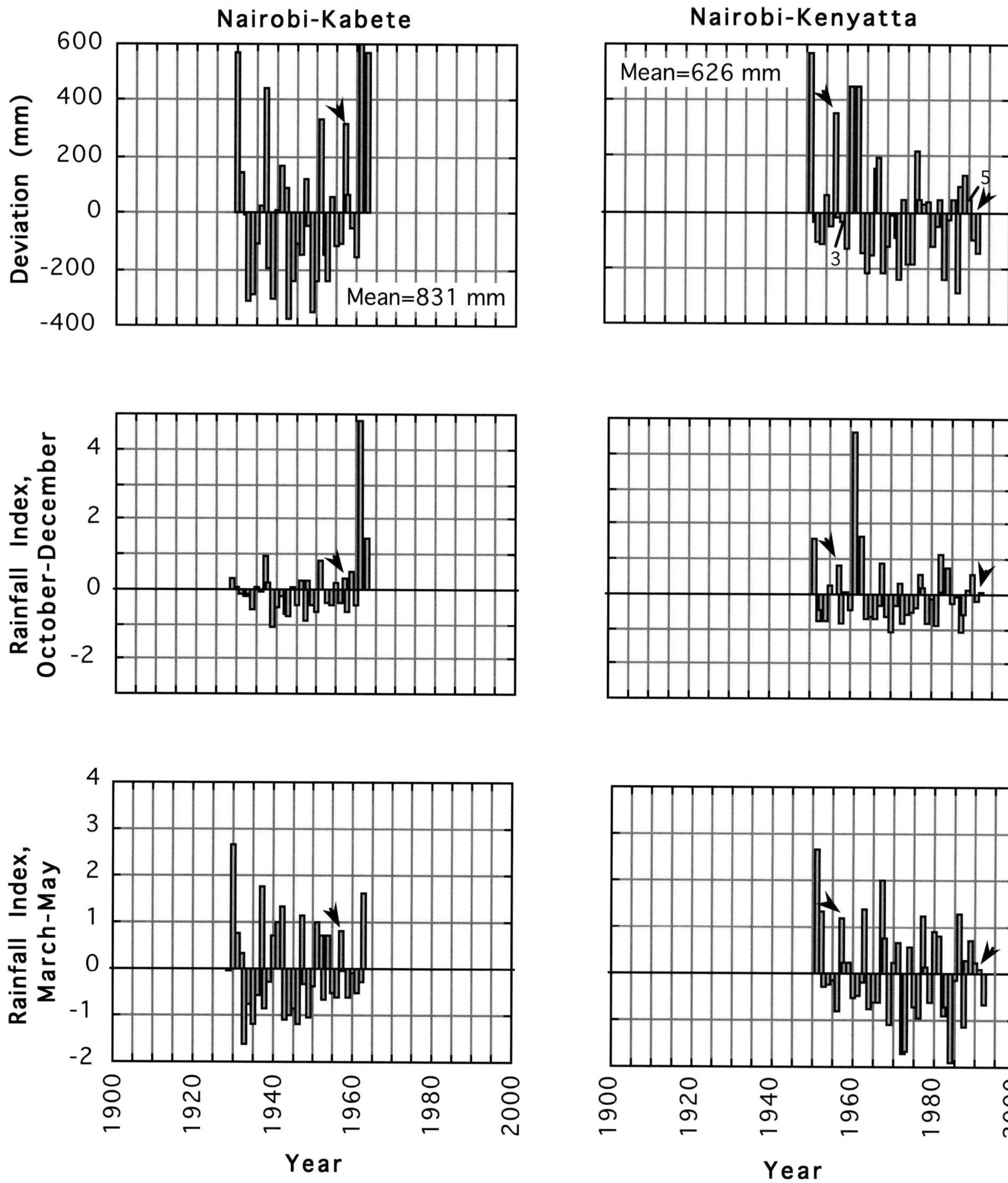
Nairobi-Kabete
Nairobi-Kenyatta
Mean=831 mm
Mean=626 mm
Deviation (mm)
Rainfall Index, October-December
Rainfall Index, March-May
600
400
200
0
-200
-400
4
2
0
-2
4
3
2
1
0
-1
-2
3
5
1900
1920
1940
1960
1980
2000
Year

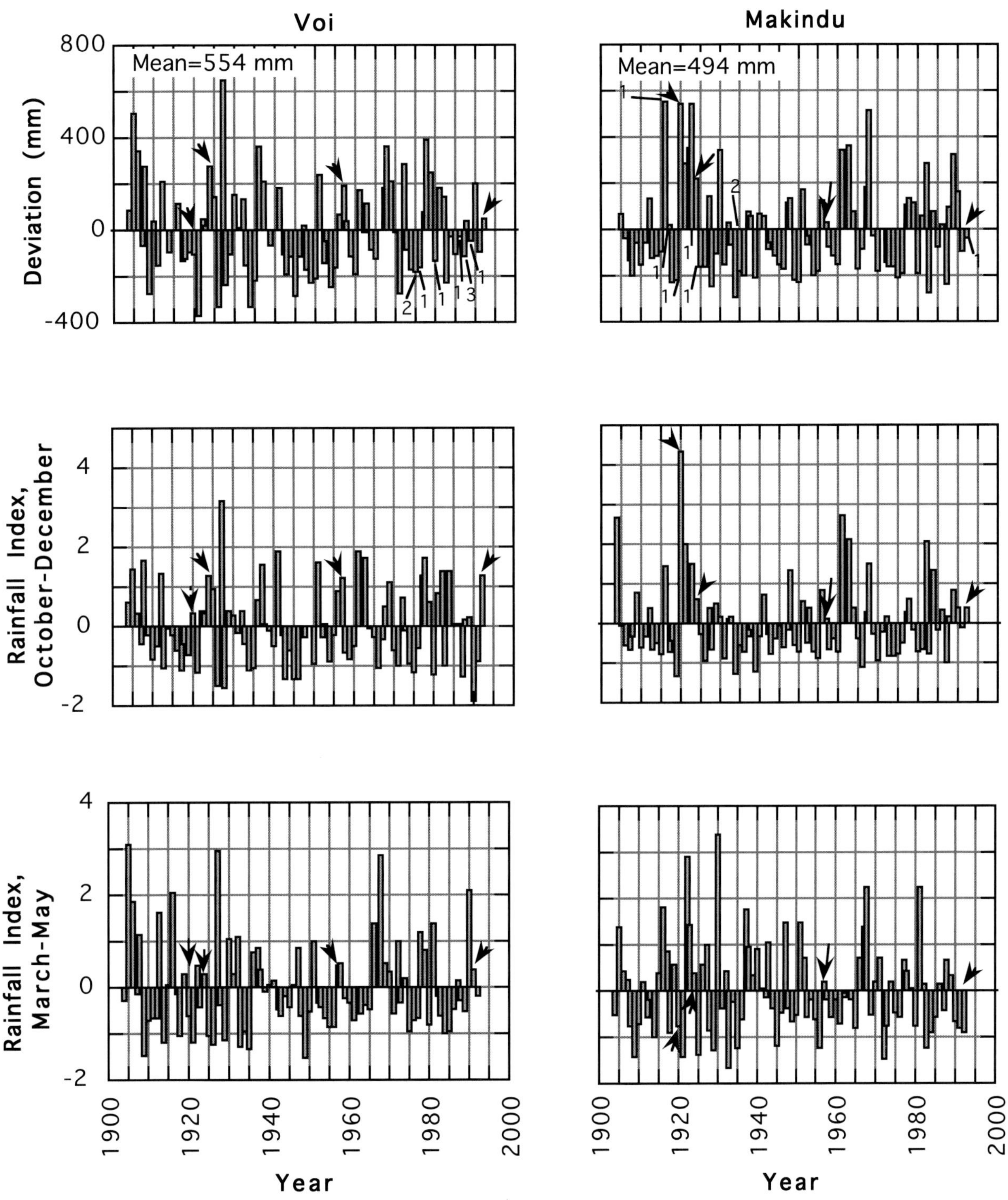
Voi
Makindu
Mean=554 mm
Mean=494 mm
Deviation (mm)
Rainfall Index, October-December
Rainfall Index, March-May
Year
Year

INDEX